HISTORY OF THE BENGAL ARTILLERY.

HISTORY OF THE ORGANIZATION, EQUIPMENT, AND WAR SERVICES OF THE REGIMENT OF BENGAL ARTILLERY,

COMPILED FROM PUBLISHED WORKS, OFFICIAL RECORDS,
AND VARIOUS PRIVATE SOURCES.

BY FRANCIS W. STUBBS,

Major, Royal (late Bengal) Artillery.

WAR SERVICES.—Vol. II.

WITH NUMEROUS MAPS AND ILLUSTRATIONS.

"ἐπεὶ μάθον ἔμμεναι ἐσθλὸς
Αἰεὶ καὶ πρώτοισι μετὰ Τρώεσσι μάχεσθαι,
Ἀρνύμενος πατρός τε μέγα κλέος ἠδ' ἐμὸν αὐτοῦ."

Henry S. King & Co., London.
1877.

CONTENTS OF VOL. II.

CHAPTER X.

PAGE

Nipál war, 1814-1816—FIRST CAMPAIGN: Plan of attack—Formation of divisions—1st, or General Marley's—2nd, or General Gillespie's—3rd, or Brigadier-General Ochterlony's—4th, or General J. S. Woods'—Captain B. Latter's column—2ND DIVISION: Advance into the Dera Dun—Assault of KALANGA—Death of General Gillespie—Second assault and failure—Evacuation of the fort—Move upon Jaitak—Unsuccessful attack—Protracted operations—3RD DIVISION: Reduction of Nálágarh—Attack on the Rámgarh heights—Flank of the Gurkhas turned—Storming of the MALAON heights—Surrender of Amar Singh and evacuation of the western hills—Services of the Artillery acknowledged—COLONEL NICOLLS' COLUMN FORMED—Captain Hearsey and Lieut.-Colonel Gardner advance into Kamáon—Reinforcement by Colonel Nicolls—Capture of Sitoli stockade—Of Almorah—Cession of Kamáon—4TH DIVISION: Advance into the Taráí—Repulse—1ST DIVISION: Occupation of the advanced posts unsupported—Their attack and capture—Gallant conduct of Lieutenant Matheson, Matross Levy, and Lascar Silári—Indecision of the general—His decision and desertion of his army—Colonel Gregory's movements in Tirhut—SECOND CAMPAIGN: Constitution of the columns—Advance of the centre column under Sir D. Ochterlony—Flank march—Elephant ladder—Action at Makwánpúr—Colonel Kelly's attack on Hariharpur—Movements of the left column—Conclusion of the war—Observations ... 1

CHAPTER XI.

SIEGE OF HATHRAS, 1817—Dyarám and Bhagwant Singh—Háthras invested—Constitution of the force—Attack upon the town—Batteries—Town evacuated—Attack upon the fort—Batteries—Parallel opened—Bombardment commenced—Explosion of a magazine—Escape of Dyarám—Surrender of neighbouring forts—Orders published—Sir John Horsford's character and death 46

CHAPTER XII.

PAGE

PINDÁRI AND MÁHRÁTÁ WAR, 1817-1819—Assembly of the army—Its constitution—Positions of the Divisions at the beginning of the war—Detail of the Centre Division grand army—Of the Right Division—Of the Left Division—Of the Reserve—Of Generals Hardyman and Toone's columns—Of the 5th Division army of the Dakhan—Positions of the Pindári *darras*—Hostile outbreak at Poonah—At Nágpur—General Doveton arrives—Battle and siege of NÁGPUR—General Hardyman attacks the enemy at Jubulpore—Holkar's army defeated at MAHIDPORE—Pursuit of the Pindáras—Advance of the 3rd, 5th, and Left Divisions—Colonel Philpot moves to cut them off from Gwalior—Right Division moves down—Left Division surprises them at Bechi Tál—General Brown attacks Rámpura—Also JÁWAD—Lieutenant Matheson—Dissolution of Pindári confederacies—Breaking up of the grand army—Left Division reinforced—Siege of DHÁMONI—Of MANDALAH—General Watson left in command — Satanwári — Siege of GARHÁKOTA — 5th Division moves on Nágpur — Movements of 2nd and 4th Divisions—Colonel Adams leaves Nágpur to follow Báji Ráo—Remarkable march of Captain Rodber's troop—Adams defeats Báji Ráo near Siuni—Good service rendered by Bengal and Madras Horse Artillery—Siege of CHÁNDA—Operations of the Reserve—Amir Khán's guns taken by Brigadier A. Knox—Brigadier Arnold reduces forts in Háriána and Bikanir—Táràgarh and other forts in Rájputána taken — Siege of ASIRGARH — Force employed — Its defences—Town occupied—Batteries opened—Explosion of an expense magazine—Engineer's report on the attack — Lower fort evacuated — Progress of the attacks east and west—Capitulation — Remarks on the war—Severe work performed by the troops—Deficiency of Artillery officers and of siege ordnance 63

CHAPTER XIII.

FIRST BURMESE WAR, 1824-1826 — Hostilities previous to war being proclaimed — Captain Timbrell's flotilla—Operations in Assam—Lieutenants Bedingfield and Burlton—Operations in Kachár—Failure of attack on Dudpátli—Lieutenants Huthwaite and Smith join—Attack on Tiláyan—Brigadier-General Shuldham's force—Artillery detail—Move upon Manipur abandoned—Disaster at Rámu—Invasion of ÁRÁKÁN—General Morison's force — Artillery with it—Advances—Four columns formed—General McBean joins—Batteries opened—Brigadier Richards storms the Árákán heights—Great mortality in the force—It is broken up—RANGOON Expedition—Detail of force employed—Artillery—Capture of Rangoon—Want of supplies—Position taken up before Rangoon—Remarks thereon—First attack on

PAGE

Kemendine—Attack on stockades north of Rangoon—Second attack on Kemendine—British position attacked—Ten stockades captured—Tenasserim provinces reduced—Dalla—Tantabeng—Kaiklo—Martaban taken—Enemy concentrate—British position invested—Severe fighting for seven days—Total repulse of the enemy—Services of Artillery officers noticed—Kok-keing—Reinforcements—Rocket Troop—Lieutenant-Colonel G. Pollock—His energy—1st Troop, 1st Brigade H.A.—Second expedition to Tantabeng—Advance from Rangoon—Sir Archibald Campbell's column—General Cotton's column—Detachment to Bassein—Sir A. Campbell countermarches on Donabyo—General Cotton attacks Panlang—But fails at Donabyo—Sir A. Campbell joins—Batteries open fire—Donabyo evacuated—Advance on Prome—Changes among Artillery—Enemy invest Prome—Tsenbike—Napádi—Captain Lumsden wounded—Honourable mention of the Artillery—Advance—Negotiations—Some Artillery officers leave—Movements in Pegu—Sitang taken—Captain Dickenson honourably mentioned—Melloon—Its storm and capture—Honourable mention of the Artillery—Advance—Action of Pagahm-Myo—Sir A. Campbell's order on the occasion—Peace 133

CHAPTER XIV.

Second Siege of Bhurtpore—Usurpation of Durjan Sál—Sir David Ochterlony's movement against Bhurtpore is countermanded—Preparations for the siege—Brigading of the army—Horse Artillery—Foot Artillery and battering train—Bhurtpore invested—Water supply for the ditches cut off—Base of attack laid—First batteries established—Desertion of an artilleryman—Cavalry and Horse Artillery on the west face—Major Whish's battery—Great mortar and breaching batteries—Breaching and bombarding commenced—Mining commenced—Trenches extended on the right—Overtures for peace—Storming postponed—Colonel Stark and Lieutenant Pennington examine right breach—Explosion in the ammunition depôt—Enemy occupy the ditch—Mine under north-east angle—Mine under long-necked bastion fired—Breach examined—Arrangements for storming—Storm and capture—Casualties in the Artillery—Ordnance captured—Officers noticed in orders ... 192

Appendix 225

Index of Officers 255

LIST OF THE MAPS, PLANS, ETC., IN VOLUME II.

	PAGE
Plan of the Fort of Kalanga	To face 3
Sketch of the Operations of the 3rd Division, from November, 1814, to May, 1815	„ 15
Sketch of the Operations against Maláon in April, 1815 ...	„ 17
Map of the Operations of Major-Generals B. Marley and J. S. Wood in 1814–15, and of Sir D. Ochterlony in 1816 ...	„ 23
Plan of the Siege and of the Town and Fort of Háthras in 1817	„ 47
Theatre of Operations in the Pindári and Máhráttá War in 1817-19	„ 65
Plan of the Battle of Nágpore, December 16, 1817	„ 75
Plan of the Attack on the Fort of Dhámoni, March 24, 1818	„ 89
Plan of Attack on the Fort of Mandalah, April, 1818 ...	„ 91
Plan of the Attack of Garhákota, October, 1818	„ 93
Plan of the Front of Attack of the Town of Chándá, May, 1818	„ 101
Plan of the Attack of Asirgárh	„ 107
Map of Burmah	„ 143
Plan of the Vicinity of Rangoon	„ 145
View of the Shwe-da-gon Pagoda, Rangoon	On 146
Position of the British Army before Rangoon, from May to December, 1804	To face 147
Bhurtpore taken, 18th January, 1826	„ 199
Plan of the Siege Operations, Bhurtpore	„ 201
North-east Angle Bastion	On 212
Long-necked Bastion	On 213
Bronze 18-pounder captured at Bhurtpore	216

HISTORY OF THE BENGAL ARTILLERY.

CHAPTER X.

Nipál war, 1814-1816—FIRST CAMPAIGN: Plan of attack—Formation of divisions—1st, or General Marley's—2nd, or General Gillespie's—3rd, or Brigadier-General Ochterlony's—4th, or General J. S. Woods'—Captain B. Latter's column—2ND DIVISION: Advance into the Dera Dun—Assault of KALANGA—Death of General Gillespie—Second assault and failure—Evacuation of the fort—Move upon Jaitak—Unsuccessful attack—Protracted operations—3RD DIVISION: Reduction of Nálágarh—Attack on the Rámgarh heights—Flank of the Gurkhas turned—Storming of the MALÁON heights—Surrender of Amar Singh and evacuation of the western hills—Services of the artillery acknowledged —COLONEL NICOLLS' COLUMN FORMED—Captain Hearsey and Lieut.-Colonel Gardner advance into Kamáon—Reinforcement by Colonel Nicolls—Capture of Sitoli stockade—Of Almorah—Cession of Kamáon—4TH DIVISION: Advance into the Tarái—Repulse—1ST DIVISION: Occupation of the advanced posts unsupported—Their attack and capture—Gallant conduct of Lieutenant Matheson, Matross Levy, and Lascar Silári—Indecision of the general—His decision and desertion of his army—Colonel Gregory's movements in Tirhut—SECOND CAMPAIGN: Constitution of the columns—Advance of the centre column under Sir D. Ochterlony—Flank march—Elephant ladder—Action at Makwánpur—Colonel Kelly's attack on Hariharpúr—Movements of the left column—Conclusion of the war—Observations.

THE NIPÁL WAR.

1814

THE Nipál war commenced towards the close of the year 1814. The plan of the Marquis of Hastings * was

* At this time his title was Earl Moira.

 B

1814 a nearly simultaneous attack [illegible] upon four principal points [illegible] Gurkhas spread [illegible] extensive frontier.* The western division [illegible] at Rupar on the [illegible] operate against [illegible] right; this was [illegible] sent towards [illegible] Division moved from [illegible] and Garhwal. The [illegible] Division from [illegible] penetrate to [illegible] On the [illegible] point, the 1st Division was [illegible] the honour of attacking Katmandu the capital. [illegible] A little afterwards [illegible] Kamaon from [illegible] the operations [illegible] though a small [illegible] of the [illegible] Captain [illegible] Latter [illegible] with a small force of [illegible] observe the frontier east [illegible] as far as the river Brahmaputra.

The [illegible] country from the river [illegible] on the east [illegible] on the west was under Nepal dominion. That part [illegible] west of the [illegible] Gogra† comprehending Kamaon, Garhwal, Sirmur [illegible] Hindur, was a late conquest of the Gurkha nation [illegible] was at the time under the command of Amar Sing Thapa, a chief noted for his patriotism and [illegible] qualities strongly [illegible] to him during the war. His [illegible] commanded under him [illegible] the Nahan district, and Captain‡ Balbhadar Sing [illegible] Dera Dun. Numerically the Gurkha forces were [illegible]

* For the numerical strength of these divisions, see Note [illegible] appendix to this chapter.

† Called in the [illegible] Ka.

‡ The Gurkhas [illegible] adopted some of the European titles of [illegible] rank, and drilled their troops after the ancient fashion, keeping time [illegible] [illegible] through the manual exercise [illegible] with as many motions as were [illegible] hundred years ago [illegible] words of command were in French [illegible]

1814 a nearly simultaneous attack by independent columns upon four principal points, with the view of making the Gurkhas spread their not very numerous forces along an extensive frontier.* The western division was formed at Rupar, on the Sutlej, to operate against the extreme right; this was the 3rd. Next towards the east, the 2nd Division moved from Sahâranpur, upon the Dera Dun, and Garhwál. The 4th Division, from Gorakhpur, was to penetrate by Butwal to Pálpa. On the east, at Dinapore, the 1st Division was formed, and to it was assigned the honour of attacking Katmandu, the capital of Nipál. A fifth afterwards invaded Kamáon from Rohilkhand, and the operations of this column, though a small one, proved of the greatest value. Captain Barré Latter, of the Rangpur battalion, with a small force of 2723 men, including artillery, observed the frontier east of the river Kosi, as far as the river Bráhmaputra.

The whole of the hill country, from the river Tista on the east to the Sutlej on the west, was under Nipalese dominion. That part of it lying west of the river Gogra,† comprehending Kamáon, Garhwál, Sirmur, and Hindur, was a late conquest of the Gurkha nation, and was at the time under the command of Amar Singh Thápa, a chief noted for his patriotism and courage, qualities strongly exhibited by him during the war. His eldest son, Ranjor Singh, commanded under him in the Náhan district, and Captain ‡ Balbhadar Singh in the Dera Dun. Numerically, the Gurkha forces were con-

* For the numerical strength of these divisions, see Note A in the appendix to this chapter.

† Called in the hills the Sarju, or Káli.

‡ The Gurkhas used to adopt some of the European titles of military rank, and drill their troops after the ancient fashion, keeping time by a flügel-man. I saw Jang Bahádur's men go through the manual exercise in May, 1853, in this manner, with as many motions as were in vogue a hundred years ago. The words of command were in French.—F. W. S.

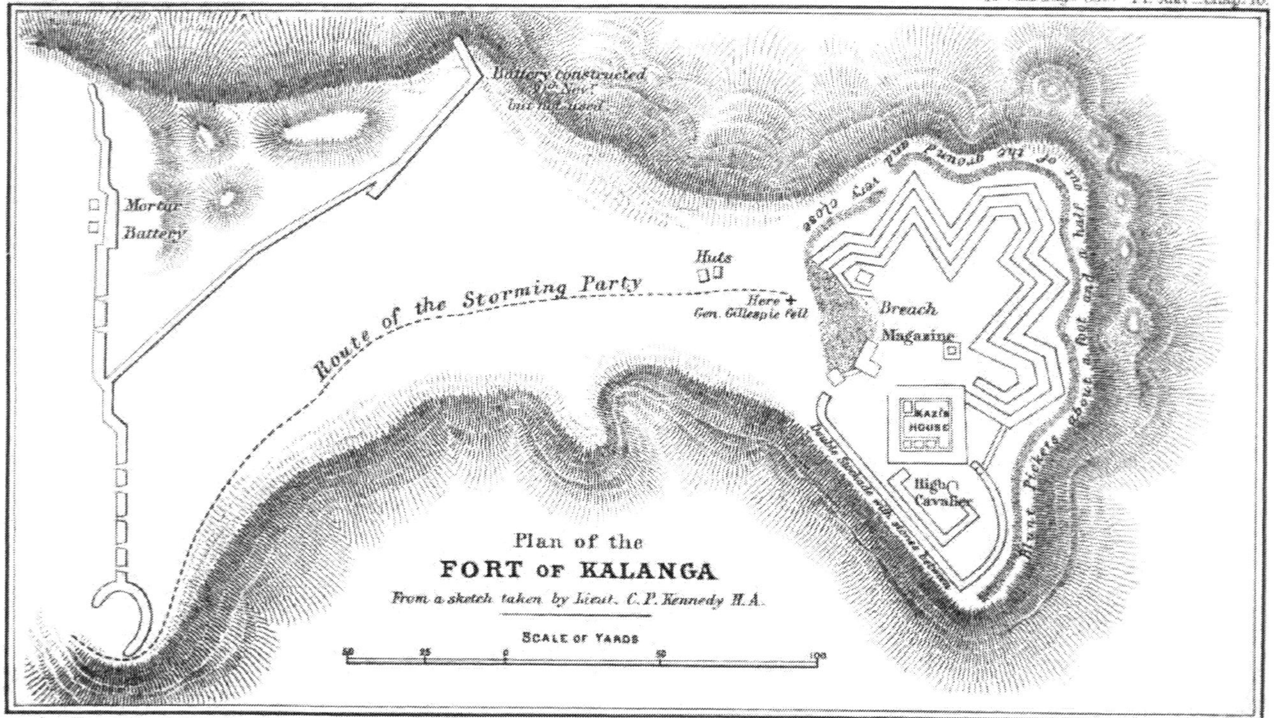

Battery constructed
but not used

Mortar
Battery

Huts

Route of the Storming Party

Here +
Gen. Gillespie fell

very close and the ground ... out of ... half a foot ... above ... Pickets ...

Breach

Magazine

KAZI'S
HOUSE

High
Cavalier

Double Stockade with stones between

Plan of the
FORT OF KALANGA
From a sketch taken by Lieut. C. P. Kennedy H. A.

SCALE OF YARDS

50 25 0 50 100

London, Henry S. King & Co. 65, Cornhill.

temptible; according to Lord Hastings, they could not at any period of the war, have amounted to 16,000 men. But they were strong, active, and intelligent mountaineers, and their courage made up for paucity of numbers.

1814

Operations of the 2nd Division, commanded by Major-General Sir Robert R. Gillespie.

The first division to move was the second. The 1st and 3rd Troops of Horse Artillery went from Meerut to join it. The officers with them were:—

1st Troop.

A-C R.H.A

Captain and Brevet-Major Gervaise Pennington.
Capt.-Lieut. William McQuhae (quarter-master to the reserve).
Lieutenant Charles Pratt Kennedy.
,, Gabriel Napier C. Campbell.
Lieut.-Fireworker Richard Scrope B. Morland.

3rd Troop.

B-C R.H.A.

Captain James H. Brooke.
Capt.-Lieutenant John Rodber.
Lieutenant Hugh L. Playfair (doing duty from 2nd Troop, and adjutant).
Lieutenant John Sconce.
,, John B. B. Luxford.

October

The division marched from Sahâranpur on the 19th of October. Two columns, under Colonel Sebright Mawbey, 53rd Regiment, and Lieut.-Colonel G. Carpenter, 17th N.I., entered the valley of the Dun by the Kheri and Timli passes, while Major-General Sir Robert Gillespie followed with the main body. Part of the horse artillery, with Rodber, Luxford, and Kennedy, accompanied the second column. Dera was occupied by Colonel Mawbey on the 22nd. About five miles from this place, on a hill of no great height, was the fort of Nálápáni, or Kalanga, where Balbhadar Singh had resolved to stand his ground. It was mostly built of stone masonry, strengthened in places with a double stockade,

1814 October

was only assailable on one small front, and contained a garrison of between 500 and 600 men. Colonel Mawbey reconnoitred the place on the 24th, and finding it too strong for a *coup de main*, returned to Dera to await Sir Robert Gillespie's arrival. This was in accordance with his instructions.[1]

On the 27th, the major-general reached Dera, where he received Colonel Mawbey's report, in which, among other officers, Captain Rodber and Lieutenants Kennedy and Luxford were singled out for praise. Determining to attack at once, he on the 29th formed five columns for the assault, and issued instructions for their guidance.* They encamped in the following order:—

First column, under Lieut.-Colonel Carpenter: Two companies 53rd Regiment, and five native infantry; 588 non-commissioned officers, rank and file.

Reserve, under Major Ludlow: 6th N.I.; 991 ditto.

Second column under Captain J. W. Fast: 17th N.I.; 351 ditto.

Third column, under Major Kelly: 7th N.I.; 519 ditto, and 20 pioneers.

Fourth column, under Captain Campbell: 6th N.I.; 273 ditto.

All except the two companies of the 53rd in the first column, and 100 dismounted dragoons in the reserve, being native infantry.

* Each column was to be preceded by twenty men with talwárs, or native swords. But the sword is no match for the Gurkha's kukri and small shield, as the number of casualties in the 8th Dragoons sufficiently proves. An amusing account of one of the encounters has been preserved,[2] which, if correct in all its details, shows that the dragoon, though a good swordsman, could gain no advantage over his antagonist until he had, by skill or luck, disarmed him of his kukri. Even then it was only by a blow of the left fist in the stomach, while the shield was raised to cover the Gurkha's head, which doubled him up, that he was able to deliver his point. General Gillespie's instructions to the assaulting columns are given in Note B.

[1] Adjutant-General to Major-General Gillespie, dated October 1st, 1814. Papers, p. 161.

[2] *East Indian United Service Journal* for April, 1836, vol. viii, p. 275.

1814 October

Next day, Lieut.-Colonel Carpenter and Major Ludlow had, without meeting any opposition, taken up a position on the level ground in front of the fort. Working parties were employed during the night under Lieutenant G. R. Blane, of the engineers, with Lieutenants Elliott and Ellis, of the pioneers, under the direction of Major Pennington, in constructing a battery nearly 600 yards from it. This was done; and by daylight ten pieces of bronze ordnance,* which had been brought up on elephants, opened their fire. The other three columns, moving by different routes, were to combine, at a given signal, with Colonel Carpenter in a simultaneous attack. This signal was the firing of five guns from the battery —an arrangement almost certain to lead to misconception and failure; not only because the batteries were at work from daylight, but also because sound among hills is so easily stopped. But there is too much reason for knowing that it was fired an hour or more before the time which had been specified.† Between the battery and the fort there was a village of thatched huts. Pioneers with scaling ladders, under Lieutenant Elliott and Ensign Ellis, accompanied the columns of Colonel Carpenter and Major Ludlow, which were now ordered forward. The dismounted dragoons of the Royal Irish Regiment were led

* Two 12-pounder guns and two 5½-inch howitzers, horse artillery; four 6-pounder battalion guns, and two 5½-inch mortars.

† All accounts agree in ascribing this change to the general's impetuosity, but not as to the ulterior cause. Prinsep (vol. i. p. 88) says that it was because the fire from the batteries had not produced as much effect as was intended. The guns being light field-pieces, this is certainly very probable. Colonel Kennedy states that a party of the garrison, who had got through the jungle upon the flank of the battery, were dispersed by a few rounds of grape, and that the general considered the opportunity a favourable one for immediate attack. The account of this affair given to me by Colonel Kennedy is of much value, as it was prepared from notes made at the time and shortly afterwards, when he was examined by a court of inquiry, which sat at Meerut to investigate the causes of the failures at Kalanga.

1814 October

by Captain N. Brutton, an officer of approved gallantry, and were first in the attack. Passing round by the left, and pressing on with their wonted alacrity, they left the rest behind, and, being thus without support, the garrison sallied out and attacked them. A close hand to hand fight ensued. The Gurkhas, with their shields and kukris, getting within the point of the sabre, had the advantage, and in a few minutes the dragoons had to give way, having lost four killed and fifty-eight wounded—Captain Brutton and two officers severely, one slightly.* The infantry part of the columns came up under a heavy fire from the walls; attempts were made to plant the ladders, but a gun placed at an open wicket of the fort, so as to enfilade this part of the front, swept down many. Ensign Ellis was killed, Lieutenant Elliott badly wounded; and they retreated. The ladders were left among the huts, which caught fire about this time.

Nothing was as yet heard of the other columns. Gillespie, on receiving the report of the failure, ordered forward three companies of the 53rd, which had just arrived in the battery, under Captain Coultman, with two horse artillery guns under Lieutenant Kennedy, to blow open the gate. Lieutenant Napier Campbell accompanied them, and the men of the 53rd manned the drag-ropes. After a difficult ascent they came to a stockade intersecting the road, within sixty or eighty yards of the nearest bastion, in getting over which the order of march was somewhat broken, and a sharp musketry fire was opened upon them. The general, accompanied by Colonel Westenra, of the dragoons, and his staff, came up with them here, and they went forward, leaving Lieutenant

* Several accounts, following Prinsep, make the general to have fallen at the head of the dragoons; but this is not the case. Colonel Kennedy's statement is borne out by the despatches, and also by other authorities. This I have followed.

1814 October

Campbell with one of the guns to cover the advance. Passing through the village, the huts still burning, and much impeded with the dead and wounded of the preceding column of attack, they came to a turn in the road in full sight of the gate of the fort, some fifty or sixty yards off—"A cut in the wall, with some loose stones piled up about four feet high, and above them two strong bars of wood." Through this was pointed a gun loaded with all manner of missiles, and a number of matchlocks were pointed at them over the wall; but now hardly a shot was fired as they advanced up the lane, partially screened by the smoke. The general and staff led, followed by the 53rd men, steadily dragging on the gun. Under his orders, Lieutenants Blane and Kennedy ran on ahead to select a position for the gun, which was brought up to about thirty or forty feet of the entrance, as far as it could go. It gave its message, but the reply was read too plainly by those who should have gone forward to the assault. They wavered, and the fire of matchlocks and arrows told with effect upon the leading subdivisions. In vain did the general repeat his orders for the men to charge. The wooden bars across the entrance were broken by the fire of the gun, and a party of stout Gurkhas rushed out and cut the only sponge-staff in two. Gillespie was frantic. Major Ludlow appeared at this juncture with several officers and sepoys, and was desired to attack to the right, where it was supposed there was an entrance; and the horse artillerymen were ordered to arm themselves with the muskets of the dead. The supply was not a scanty one. When this was done, the general, with his sword in one hand and a double-barrelled pistol in the other, turned to Lieutenant Kennedy and the rest, exclaiming, "Come on, my lads; now, Charles, for the honour of the County Down!"

1814 October

Only a pace or two forward, and he fell with a bullet through his heart. The body was taken to the rear by Sergeant Hamilton (another County Down man), Sergeant Moseley, of the dragoons, and some horse artillerymen. The order to retire was given, but in the confusion it was not heard in front, and gun and gunners were nearly lost. Captain Campbell fortunately appeared with his column at this time, and assisted in getting it away and covering the retreat. So failed the assault; a proof, if one were needed, that, in such cases, to hesitate is to throw away all chances of success, and, if the paradox may be allowed in a limited sense, that it is better to do wrong than to do nothing.

November

Colonel Mawbey, on whom the command devolved, fell back again to Dera till the 23rd of November, when a train of four 18-pounder guns and two 8-inch mortars, along with the head-quarters of the 5th and a detachment of the 6th Company, 3rd Battalion Artillery, under Captain-Lieutenant W. Battine,* arrived from Delhi. The following officers were with these companies:—

A-8 R.A.
B-19 R.A.

Capt.-Lieutenant ...	William Battine.
Lieutenant	Charles Cornwallis Chesney, adjutant.
"	James Tennant.
"	Theodore Lyons.
Lieutenant-Fireworker...	Charles Smith.
"	Charles G. Dixon.

Lieutenant Edward Hall had been ordered up, a short time before, with some small mountain mortars and howitzers. Captain W. H. Carmichael Smyth, of the engineers, was sent up to superintend the operations.

* Captain Battine had been commanding the 4th Division of Field Artillery at Rewári, and it had been intended that he should assume command of the artillery with the 2nd Division. The 4$\frac{2}{5}$-inch howitzers and mortars were supplied with carriages and beds adapted for transport by coolies.—Adjutant-General to Major-General Gillespie. Papers p. 165. See Note D in the Appendix.

1814 November

The 18-pounders were placed in battery only 200 yards from the walls; and on the 27th, the breach being reported quite practicable, the assault was ordered. Again the report chronicled a failure. Only Lieutenant Harrington, of the 53rd, with a few men, ascended the breach, and he was killed. The storming party was reinforced; and Lieutenant Edward Hall, with a 12-pounder, and Lieutenant John B. B. Luxford, with a 5½-inch howitzer, were sent forward. The latter officer, while gallantly endeavouring to drag his howitzer over the breach, was mortally wounded, and died on the 29th. The loss on this day was 4 English officers, 17 Europeans and natives killed; 7 English officers, 215 Europeans, and 221 natives wounded—of which the horse artillery had 3 killed, 9 wounded, and the foot artillery, 7 wounded.

Colonel Mawbey, then under orders from Brigadier-General Ochterlony, in whose command this division was temporarily included, cut off the supplies and water from the fort; and on the night of the 30th, the remains of the garrison, now reduced to less than 150 * men (inclusive of some women and children), evacuated it, and made their way with some loss through the investing force, leaving behind them their dead and wounded. Within the fort the scene, which no one cared to look upon a second time, sufficiently attested the stubbornness of the defence. It had cost the assailants, in two assaults, 75 killed and 680 wounded. †

* Colonel Mawbey reports seventy, but he does not take into account the loss suffered by them in retreat, or the women and children.

† Including two who do not appear in the numerical statement, viz. Lieutenant-Colonel Westenra, 8th Royal Irish Dragoons, and Captain J. S. Byers, Royal Artillery, the major-general's aide-de-camp.

Lieutenant Kennedy next morning rode his horse, an Arab, over the breach into the fort, so that it must have been perfectly practicable. Stout hearts are a better defence than stone walls.

It may here be mentioned that the brave Balbhadar Singh was sub-

1814 December

The following is extracted from the orders issued by Colonel Mawbey on this occasion:—

"Camp near Kalanga, Saturday, 1st December, 1814.

". . . . The possession of the hill by the advanced guard, on the 21st ultimo, under Major Richards, and the eligible spot fixed upon by Captain Smyth, of the engineers, and the correctness with which Lieutenant C. P. Kennedy, of the horse artillery, availed himself of favourable ground for the howitzers, which quickly drove the enemy from their advanced stockades . . . . claims Colonel Mawbey's particular notice.

"The exertions of Major Pennington and the officers and men of the artillery in getting the guns and ammunition up the hills, and their laborious duty in the battery, the effect of which is perceptible on examining the fort, speak so fully for themselves, that anything Colonel Mawbey could say would be superfluous."

The plans of the Governor-General had now been so far altered that this division, instead of invading the hill district of Garhwál, proceeded against Náhan,* near to which Ranjor Singh was posted. Major-General Gabriel Martindell had been appointed to the command, and joined on the 20th of December. The horse artillery had not been found adapted for this hill warfare, in which they could only act as dismounted gunners, and they had already left for Meerut. Lieut.-Colonel Carpenter was posted at Kálsi, at the north-west extremity of the Dun, and Colonel Mawbey had also occupied Barát, an elevated post of the ridge to the north-east, about seven miles distant in a direct line, which the Gurkhas by this move were obliged to abandon. The communications of Amar Singh and Ranjor Singh with

sequently obliged to leave Nipál, his life having become forfeited to the vengeance of a countryman, whose wife he had seduced. He entered the service of Ranjit Singh of Lahore, and died in March, 1824, fighting under the Máhárájá against the Khatak Patháns in a manner worthy of his former reputation.—"History of the Punjab" (edited by Prinsep), vol. ii. p. 73. London, 1846.

* It was Ochterlony who suggested this move.

the kingdom of Nipál were now limited to very difficult routes, close under the snowy ranges.

1814 December

The division took up its position below Náhan on the 19th,* and occupied that place on the 24th. Ranjor Singh had, under orders from his chief, taken post at Jaitak, on a high range not more than four miles north-east from the Barát ridge, but nearly 2000 feet higher. After an examination of the position, two columns of attack were formed. One, under Major W. Richards,† 1—13th N.I., consisted of the light company 53rd Regiment, 250 rank and file of the light N.I. battalion, 273 of the 1st Battalion 13th N.I., and 50 pioneers. The other, under Major J. Ludlow, was composed of the grenadiers of the 53rd, 250 rank and file light N.I., and 588 of the 1st Battalion 6th N.I., and a like number of pioneers. The first column was to proceed by a detour, and take possession of a hill, since known as the Peacock Hill, north of the Jaitak peak; the second was to take the latter in front. A 6-pounder and a 5½-inch howitzer, carried on elephants, were sent with each. Lieutenant C. Smith accompanied Major Richards, and Lieutenant T. Lyons, Major Ludlow. Their instructions were to endeavour to cut off the enemy's water supply, and intercept their communications; and Major Ludlow was, on attaining the summit, to endeavour to shell out the garrison before making an assault.[1]

The columns commenced their march during the night of the 26th. Major Richards moved off first, having

* Lieut.-Fireworker E. P. Gowan joined the 2nd Division about this time.

† Afterwards General Sir William Richards, K.C.B. He died at Naini Tál, where he had lived for some years, on the 1st of November, 1861.

[1] Major-General Martindell to Adjutant-General, 27th December. Papers, p. 503.

1814 December

the longest way to go, and took possession of the ground he was ordered to occupy. Major Ludlow fell in about 3 a.m. with the Gurkha advanced picquet, which retired up the hill upon the ruined village and temple of Jamta, where was a second post, and a little further on a stockade. The difficult paths they had followed, and the pace kept up by the infantry, caused the guns to be left in the rear; but the grenadiers of the 53rd, anxious, it would appear, to atone for the failures at Kalanga, pushed forward before the native infantry, still filing up to the position at Jamta, could form up. So that when, close under the stockade, they were received with a sharp fire on both flanks and in rear, their ardour received a sudden check; and, the Gurkhas issuing forth from their position in front, with shield, sword, and kukri, unsupported, they gave way before the dreaded charge of these brave mountaineers in spite of all efforts to rally them. The native infantry, seeing the disorderly retreat of the Europeans, caught the contagion—their officers were too few to hold them in check and present a firm front—and the panic spreading, as it ever does, in extent, rapidly carried the whole back to camp.

This placed Major Richards' column in a very critical position. He kept his ground, however, throughout the whole day, and sent to Major-General Martindell for reinforcements of men, and particularly of ammunition, which had run very low in repulsing the reiterated attacks of the enemy. Two hours after sunset, however, a positive order to retire arrived, and under the circumstances there was no alternative left. That this retreat was successfully conducted, in the darkness of the night, along a narrow and dangerous pathway, and closely followed by the enemy, reflected the highest praise on the ability of the leader and the steadiness of his troops. And the heroic devotion of Lieutenant Thackeray, who, with a company of the light N.I. battalion, kept the Gurkhas in check and mainly contributed to the safety of the column, by repeatedly charging wherever he had a foe to meet, should not be omitted, though he did not belong to the arm whose services this work is intended to record. He, with Ensigns Wilson and Stalkart, fell in the execution of their duty worthily performed.

1814 December

This untoward failure induced General Martindell to abstain from further active operations till reinforcements, drawn with difficulty from other points, were received.* In February, he occupied a position on the ridge higher up than that from which Major Ludlow had been driven; and his 18-pounder guns were, with infinite labour, and to the astonishment of the enemy, dragged up to a spot called the Black Hill, from whence they opened fire on the 17th of March. Another battery, more advanced, was erected on the 20th, the fire of which in one day levelled entirely the first stockade. Instead, however, of following up this success by an assault, General Martindell again changed his tactics, and threw away all his labour by turning his operations into a blockade—a decision which greatly mortified the Governor-General, already sufficiently disappointed in the utter failure of his plans as regarded the 1st and 4th Divisions. This blockade continued, however, to be maintained till the month of May, when the surrender of Amar Singh and the annexation of Kamáon placed all the hill country west of the Káli in British hands.

1815 January

February

March

May

* The ordnance now consisted of four 18-pounders and six 6-pounder guns; two 8-inch, two 5½-inch, and two 4⅖-inch mortars; two 5½-inch, and two 4⅖-inch howitzers.—"Lord Moira's Secret Letter," Papers, p. 708.

OPERATIONS OF THE 3RD DIVISION, COMMANDED BY MAJOR-GENERAL DAVID OCHTERLONY.

1814

November

3-22 R.A.

Within a few days after Gillespie's force had entered the Dun, this division moved from Rupar, and on the 2nd of November took up a position in front of the fort of Nálágarh. The 4th Company, 3rd Battalion of Artillery was attached to it, with ordnance amounting, at first, to two 18-pounders and ten 6-pounder guns, two 5½-inch mortars, and two 5½-inch howitzers. Major Alexander Macleod commanded, and Lieutenant Kenneth Cruikshank was adjutant and quarter-master. The officers with this company were:—

Captain	N. S. Webb.
Lieutenant	Charles Graham.
,, ...	George Brooke.
,, ...	Thomas Timbrell.
,, ...	Kender Mason.*
Lieut.-Fireworker	Edward P. Gowan.
,, ...	John Cartwright.

Lieutenant Lawtie, of the engineers, was immediately sent to select a spot for a battery, which being approved of by Major Macleod, the latter was able to open fire early on the morning of the 4th from two 18 and two 6-pounders; and next day, the fort, with its outlying post of Táràgarh, surrendered. The thanks of Lord Moira, as commander-in-chief, were accorded to Major Macleod and the officers and men of the artillery on this occasion.[1]

Towering above the British camp at Nálágarh rose the heights on which Amar Singh, the Gurkha chief, had

* This officer may not have been present at all with his company. He had not joined it up to the 1st of January, 1815, probably owing to the state of his health, which was very infirm.

[1] Adjutant-General to Colonel Ochterlony, dated 18th December. Papers, p. 455.

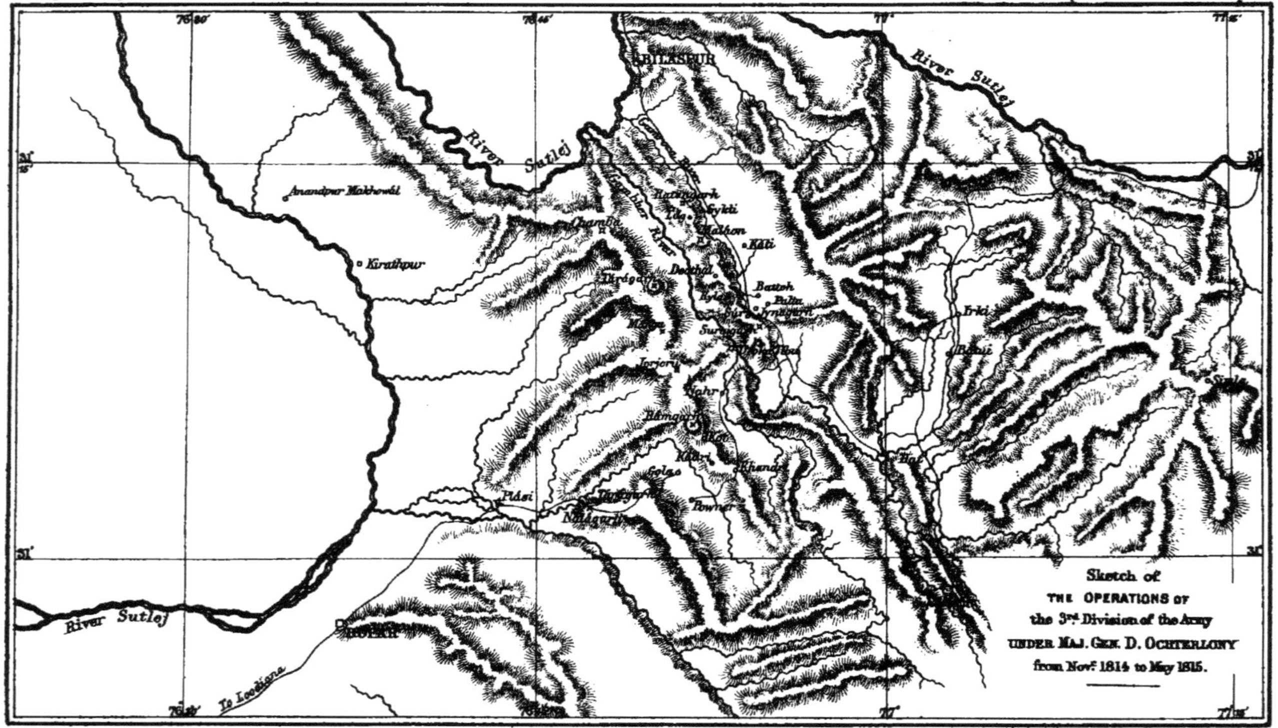

Sketch of
THE OPERATIONS OF
the 3rd Division of the Army
UNDER MAJ. GEN. D. OCHTERLONY
from Novr. 1814 to May 1815.

River Sutlej
River Sutlej
River Sutlej
BILASPUR
Anandpur Makhowál
Kírathpur
Batteh
Palta
Erki
Khandi
Powner
Golas
Pláśi
Nálágarh
RUPAR
To Loodiana

London: Henry S. King & Co. 65, Cornhill.

Edwd. Weller, Litho.

posted himself, his left resting upon Kot, a high but not fortified hill; the right of his main position on Rámgarh, in the centre of the ridge which extended as far as the river Sutlej, crowned with strong forts and abounding in formidable positions. Behind, and nearly parallel to this range, ran another, on which stood the forts Ratangarh and Maláon and Surajgarh, between the Gambher and Gamrora, small confluents of the Sutlej. The Rájá of Bilåspur, whose district lay on both sides of the latter, was as an ally of the greatest importance to Amar Singh, as nearly all the supplies for his army were drawn from thence. Irki, which had been his former head-quarters, supplied the rest, and his communications with the other positions were by this route. Both lines, therefore, were important. Amar Singh's position at Rámgarh proving to be unassailable from the front, an attempt was made to attack and turn the left, with the intention of first carrying the hill of Kot, which was occupied, but not stockaded, and from thence the rest of the posts upon the ridge successively.

1814 November

See Plate XXXII.

Lieut.-Colonel Thompson, with the reserve, was, on the 13th of November, ordered to occupy the village of Kádri. The road to this post being impracticable, it was necessary first to construct one for elephants and coolies, to convey two 6-pounders, two mortars, and two howitzers, which Major Macleod succeeded in getting, along with the ammunition and stores, to that place on the 20th. This, even after a road had with immense labour been constructed, was two days' hard work. The battery, when it opened fire, was too distant to produce any good effect, and an advanced position was being taken up for another, when the enemy attacked in great force, and the party sent forward was repulsed with considerable loss.

26th November

1814 November

Rightly judging, therefore, with the late instances of failure at Kalanga and Jaitak, that no further risk should be run, General Ochterlony changed his point of attack, and occupied himself, meanwhile, in straitening the enemy as much as possible, and improving the roads for guns, as well as embodying irregular troops, which, to the amount of 4463 men, he was authorized to add to his force. He had, on the 24th of November, moved his head-quarters round to Nahr,* on the north-north-east side of the Rámgarh ridge. On the 27th of December he was joined by the 2nd Battalion 7th N.I.; and his ordnance, with the *personnel*, now amounted to the detail given in the margin. Lieutenants J. Tennant, C. Graham, and T. Timbrell joined with two 18-pounder guns and four mountain pieces.

December

18-pounders	2
6-pounders	10
5½-inch mortars	2
4⅖-inch mortars	4
5½-inch howitzers ...	2
4⅖-inch howitzers ...	2
European and native artillerymen, lascars, and drivers	1061

The general's object now was, by placing himself between the principal position of the enemy and Biláspur, to cut off their supplies. He gained over the Rájá of Plási and Hindur, and with his aid constructed a road by Khandni to Nahr. Amar Singh promptly met this by abandoning all his posts on that ridge to the left of Rámgarh, and changing front to the rear upon that point, which he still retained as his right. General Ochterlony therefore marched, on the 16th of January, with the reserve, across the Gambher river to a position on the road to Irki, and near the southern extremity of the Maláon range. Lieut.-Colonel Cooper was left with a battalion and

1815 January

* Spelt Nori on the Indian atlas, sheet No. 47. It is not easy to identify many of the places mentioned in the despatches and accounts of this war.

(To face Page 371) Pl. XXXIII _ Chap 10.

SKETCH OF MAJOR GENERAL D. OCHTERLONY'S OPERATIONS AGAINST MALÁON IN APRIL 1815.

a. Route of Lieut. Fleming on Ryla on the night of the 14th.

b. Do. of Lieut. Lidlie & Capt. Hamilton to Jynagar on the 14th.

c. Do. of do. from Jynagar on Ryla on the 15th.

d. Do. of Major Innes from Hd. Qrs. on Ryla.

e. Do. of Lt. Col. Thompson from do. on Deonthal.

f. Do. of Capt. Lawrie from Káli on do.

g. Route of Capt. Bowyer from Káli on Maláon.

h. Do. of Capt. Showers from Ratangarh on do.

i. Stone Redoubts.

k. Górkha Stockades.

l. Capt. Stewart's Post at Lág Hill.

From Official Records & Plan.

London: Henry S. King & Co. 65, Cornhill.

Edwd. Weller, Litho.

the battering train at the former post of Nahr. Colonel Arnold, with the rest of the brigade, had orders to move either upon Biláspur or Maláon, or upon Mángu ka Dhár, according to circumstances. Amar Singh, as the general had foreseen, in consequence of this, quitted his position for Maláon, leaving only small garrisons in Rámgarh and the other posts upon that range. Colonel Arnold then proceeded to occupy the stockades evacuated, an operation retarded by the inclemency of the weather and the difficulties of the ground. This portion of the force then took post at Ratangarh, between the Maláon and Biláspur, thus accomplishing the object of turning the enemy's flank and intercepting his supplies.

1815 January

February

Meanwhile, Lieut-Colonel Cooper, as soon as these movements permitted, marched from Nahr, and ascended the Rámgarh ridge. Captain Webb, with his company, or a part of it, accompanied this detachment. The 18-pounders under Lieutenant Tennant were brought up with incredible labour, and a battery was opened upon Rámgarh, which soon surrendered. Jorjori capitulated at the same time. Tárágarh (a second place of that name) on the 11th, and Chamba on the 16th of March, were breached and taken. All were stone forts. The posts on this ridge having been thus successively reduced, the detachment took up the position which had been assigned to it before Maláon, on the 1st of April.

March

The accompanying sketch will give some idea of the different Gurkha posts, from Surajgarh fort on the south-east, to that on the Maláon on the north-west. General Ochterlony, having obtained from Lieutenant Lawtie, of the engineers, an active and highly intelligent officer, the necessary information as to locality and ground for the outlines of his plan, ordered the attack to take place on the 15th of April. Two objective points were selected—

See PLATE XXXIII.

April

1815 April

Raila, between Dáb and the stockade on the first Deonthal hill, and the second hill called Deonthal—while a diversion was to be made by an attack from the north-west and north-east, upon Maláon itself. In all, seven columns were told off, as follows:—

1. One, commanded by Lieutenant Fleming, to march from Palta by night, and, on reaching Raila, to show a light as a signal to the other columns.

2. Lieutenant Lidlie with two companies and two 6-pounders, and Captain Hamilton with another detachment, to move from Lág hill on the 14th, and unite at Jainagar, whence, on the morning of the 15th, they were to move upon Raila.

3. Major Innes, with a grenadier battalion and two 6-pounders, to move from head-quarters at Battoh, on the morning of the 15th, upon Raila.

4. Lieut.-Colonel Thompson, with the 2nd Battalion 3rd N.I., and two 6-pounders under Lieutenant J. Cartwright, to move from Battoh, on the morning of the 15th, upon the second Deonthal.

5. Captain Lawrie to move at the same time from Káli, upon the same point.

6. Captain Bowyer to leave Káli at the same time as Captain Lawrie, and, after crossing the Gamrora river, to diverge to the right and endeavour to penetrate the Maláon cantonment from the rear.

7. Captain Showers to move from Colonel Arnold's post at Ratangarh, at the same hour, and endeavour to penetrate the Maláon cantonment, between the Kakri stockade and the fort.

Besides these, Lieutenant Dunbar, with a small body of regular and irregular sepoys, was intended to act as a support for the two last detachments.

Raila was occupied without much opposition. Nearly

1815 April

the whole of the fighting fell to the columns under Lieut.-Colonel Thompson and Captain Showers, the casualties in each numbering 216 and 103 respectively. Captain Showers was killed, and his men, disheartened, retired precipitately on the Lág village. The artillery at Ratangarh arrested the pursuers, and they were enabled to re-form and act offensively. The position at Deonthal was carried after a severe struggle; and next morning, the Gurkhas came on again, to the number of nearly 2000, under Bhagti Thápá, and fought with their usual daring and perseverance, directing their efforts principally against the guns. The artillerymen were nearly all wounded * at them, and Lieut.-Fireworker Cartwright, with the only matross left undisabled, continued to serve one, while Lieutenant Hutchinson, assistant engineer, and Lieutenant Armstrong, with two sergeants of the pioneers, manned the other. The contest lasted for more than two hours. Amar Singh stood, with his younger son and colours, close by, encouraging his men; but a small reinforcement arrived from Raila, and the death of Bhagti Thápá terminated an attack, in which the steady, unflinching behaviour of the British troops made it no discredit for the Gurkhas to have failed. It is to be noticed also, that this success was won with a much smaller and more equal proportion of men than had before been matched against the enemy.

May

Major-General Ochterlony, without resting to think of his success, immediately commenced to construct a road for heavy artillery to Deonthal. On the 9th of May, a battery was completed near a redoubt under Naráyan Kot; next day, twelve pieces—12-pounders

* Six matrosses, three lascars, and one driver were wounded; none killed. One matross and one driver of this detail had been wounded the day before. Lieutenant Cartwright's detail for two guns, therefore, consisted only of eight matrosses, beside lascars; no non-commissioned officers.

1815 and howitzers—opened fire, and on the following day the 18-pounders, which had been brought up from Rámgarh, were placed in position. Amar Singh had steadfastly refused to listen to the solicitations of his followers, and submit. But now the news of the fall of Kamáon, and the desertions which, since the 16th of April, had reduced his followers to not more than 200, left him no choice, and a convention was entered into, by which he and his son Káji Ranjor Singh Thápá, then defending the fort of Jaitak against Major-General Martindell's force, were to be allowed to return with their followers and private property to Nipál, leaving all the rest of the hill country, from the Káli to the Sutlej, in the hands of the British.

The services of the artillery in their operations were thus acknowledged:—

"In operations of the nature of those conducted on the Maláon range, the services of the engineer, artillery, and pioneer departments are of a peculiarly arduous nature, and of proportionate value. His Excellency recognizes with unfeigned satisfaction, throughout the whole course of these operations, the same zeal, ability, and perseverance which have characterized these branches of the service wherever they have been called into activity in the present war; and in no situation have these qualities been more conspicuous than with Major-General Ochterlony's division.

". . . . His Excellency has equally to offer the tribute of his applause to the intrepid gallantry of Lieut.-Fireworker Cartwright, who, when the desperate perseverance of the enemy had left him with only one man unwounded, with that one man secured his gun; the other being manned with equal zeal and valour by Lieutenants Armstrong and Hutchinson, and two sergeants of the pioneers."[1]

"I feel greatly indebted to Major Macleod and the officers of artillery, in the judgment shown in the disposition of the ordnance, and for the skill evinced by their fire in a very early impression on the walls."[2]

---

[1] G. O. C. C. Head-Quarters, Fatehgarh, 26th April, 1815.

[2] Major-General Ochterlony to Adjutant-General, from Maláon, no date.

"The unwearied alacrity, the labour, the conspicuous gallantry, and the skill displayed by the whole of the artillery, engineer, and pioneer departments throughout the course of the service . . . . have been pointed out to the special notice of the Governor-General; and his Excellency accordingly professes his earnest sense of the meritorious conduct exhibited by Major Macleod, commanding the artillery, by Captain Webb, of the same corps . . . . as well as by all the officers belonging to those departments during the campaign."[1]

1815

OPERATIONS IN KAMÁON.

A movement upon Kamáon was contemplated from the first by Lord Moira, but the number of regiments required for the other points of attack left him almost without the means of making it. As events proceeded, however, and his information regarding that province became more precise, he determined to make an effort. The paucity of Gurkha troops there invited it. Two officers of the "country service," Captain H. Y. Hearsey and Lieut.-Colonel W. L. Gardner,* were directed to raise a small force of irregular infantry, and, with a detail of native artillery from Bareilly, with four 6-pounder guns, to march by different routes upon Almorah.

February

Captain Hearsey, in February, moved up along the river Gogra or Sarju, by Barmdeo and Fort Timla, and occupied Champáwat, the ancient capital of Káli Kamáon; but here he was attacked by Hasti Dal Sáh, the Gurkha commander from Doti, and his men failing him, he was taken prisoner.

March

April

February

Lieut.-Colonel Gardner was more successful. He advanced by Chilkia, along the course of the river Kosilla, dislodging the enemy from different positions, and securing his own with considerable skill, till he had

* These officers had come over from the Máhrátá service when war was declared with Sindiah in 1803.

[1] G. O. C. C. Fatehgarh, 21st May, 1815.

1815 established himself on the Chaumunh hill, about eighteen miles west, in a direct line from Almorah. From this

March place, after being joined on the 22nd of March by a further reinforcement of irregulars, he advanced upon Katarmal, which the enemy vacated; and having thus turned their position at Almorah, he employed himself in improving his means of defence in the abandoned stockades, until the force, which was coming up to join him, should arrive.

This force Lord Moira, now fully alive to the importance of the undertaking, placed under command of Colonel Jasper Nicolls, H.M.'s 14th Regiment, and then quarter-master general of H.M.'s forces in India. Along with it was sent a detail of artillerymen* under Lieutenants C. H. Bell and R. B. Wilson, and the following ordnance:—Four 6-pounder and two 12-pounder guns; two 8-inch and two 4⅗-inch mortars. Its leading

April battalion reached Katarmal on the 8th of April, and a day or two afterwards an artillery salute at Almorah told the English camp that Hasti Dal had arrived there. On the 23rd Major R. Patton, 5th N.I., sent forward with a detachment, came up with him at Ganna Náth, about fifteen miles north-east of Almorah, as he was moving round apparently to gain Nicolls' rear, and completely defeated him. The chief himself and his next in command were mortally wounded. A 6-pounder and a 4⅗-inch mortar accompanied this detachment, but the elephant with the former was sent back to camp, its pace being slow.

On the 25th the stockades at Sitoli were successfully carried, and the post of Hari Dungri, overlooking Almorah, was occupied. It was taken and retaken

* It does not appear what companies these were, but they were probably details of golandáz from Rohilkhand.

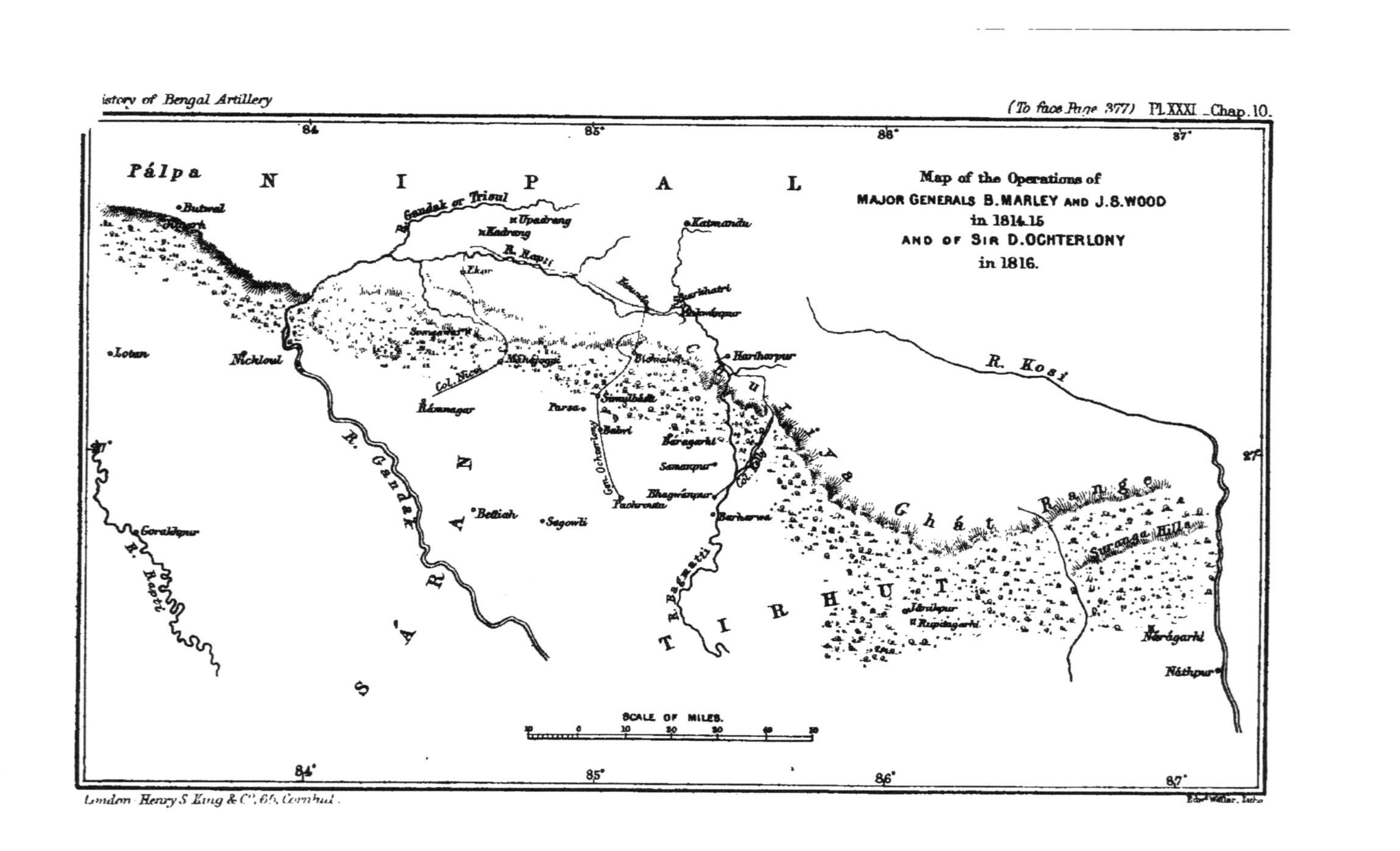

London. Henry S. King & Co. 65, Cornhill. Edwd Weller, Litho.

during the course of the night. The smaller forts opened their fire on the Káli Mandi * fort at 6 o'clock the same evening, and one of the 8-inch mortars at midnight; and next morning the advanced posts were pushed up to within seventy yards of the fort, and a few large shells, thrown into it by Lieutenant C. H. Bell, drove a portion of the garrison out. At 9 o'clock p.m. a flag of truce arrived; and next day a convention was signed, by which the whole province of Kamáon was ceded to the British, and the enemy's troops were allowed to return to Nipál.

1815 April

For their services, both Lieutenants Bell and Wilson were mentioned in terms of strong commendation by Colonel Nicolls in his report, and by the Governor-General.[1]

Thus this expedition, undertaken as quite a subordinate measure, bore the most fruit. Ochterlony had indeed left Amar Singh no resource but submission. The cession of Kamáon, however, in the language of the Gurkha chief, "broke the camel's back;" it brought the Nipalese authorities to agree to the draft of a treaty which Lord Moira proposed as the basis of peace. Let us now see what the 4th and 1st Divisions were about.

Operations of the 4th Division, commanded by Major-General John S. Wood.

The objects intended to be accomplished by this division were to recover the Tarái † of Butwal and

1814

See Plate XXXI.

* This is now called the old fort, and is situated at the northern end of the town. There was another fort called Lál Mandi, now the site of a hexagonal structure of masonry called Fort Moira, with bastions and a ditch, but no water for the garrison.

† A term usually assigned to the belt of forest skirting the foot of the hills from the Dera Dun to Bengal, varying from five to ten miles in

[1] G. G. O. Fatehgarh, 3rd May, 1815.

1815 April

Sheorāj, and to create a diversion in favour of the 1st Division, penetrating, if possible, the hills as far as Pálpa and Tonsain, a principal station and depôt of the enemy in that quarter. The 5th Company, 2nd Battalion of Artillery accompanied.* The following officers were present with it at the time:—

4-23 R.A.

Captain-Lieutenant John McDowell, commanding.
Lieutenant Thomas Croxton, doing duty from 2nd Company, 2nd Battalion.
Lieut.-Fireworker Gavin R. Crawfurd.

The ordnance attached at first were four 6-pounder and three 3-pounder guns, three 4⅖-inch mortars, and 4⅖-inch howitzers, subsequently increased by two 18-pounder and four 6-pounder guns, the 3-pounders being then withdrawn.

December

Owing to the difficulty of collecting carriage and bearers, this division did not move from Gorakhpur, where it assembled, till late in December, when it marched into the Tarái; and the general, halting at Simra, began to collect information. The swamps of the Tarái, owing to malaria, are not safe before the middle of November, and in March become again too unhealthy for any one to remain or sleep there, unless raised above the ground out of the reach of miasmatic influences. But this consideration did not hasten Major-General J. S. Wood's proceedings. Having heard that the enemy had taken post at Butwal, at the foot of the pass there, leading into

1815 January

width, and properly called the "Bhábar." Tarái (from the Persian word *tar*, moist) is the outlying belt of swamp, for the most part running along the lower margin of the forest. All this tract abounds in the larger game, as every sportsman in India knows. The Sál tree, so extensively used in making gun carriages (*Shorea robusta*), and a magnificent specimen of forest vegetation, is distinctive of this region.

7-23 R.A.

* The 3rd Company, 3rd Battalion, commanded by Captain George Pollock, was subsequently ordered up from Benares to join, but had nothing to do.

1815 January

the hills, he, on the 3rd of January, directed his march towards that point; but a report of the difficulty of the direct route, and the advice of one Kákannaddi Tewári, a Brahmin servant to the old Rájá of Pálpa, who offered himself as a guide, induced him to change his plans, and attack the stockade of Jitgarh, one mile west of Butwal. Led by this worthy, the general and his staff came upon the stockade before they were aware of it; and while the guide, after pointing it out, employed himself in disappearing from the scene, the enemy opened a smart fire upon them. The brigade-major, Captain Hiatt, and Lieutenant Morrison, field engineer, were both wounded, the latter mortally. The stockade, however, was carried in good style by Colonel Hardyman, at the head of his regiment, the 17th Foot. The other posts near it were also taken; but as the hill in rear was still held, the major-general, thinking the place untenable, "determined to stop the fruitless waste of lives by sounding the retreat."[1] Among others, Captain-Lieutenant McDowell was severely wounded. Conduct like this, if it did not increase the confidence of the British troops in their leader, did that of the enemy; Major-General J. S. Wood henceforward confined himself to purely defensive measures, although his force largely outnumbered the enemy. With the exception of constructing some works to defend itself at Lotan, and a fruitless march to Nichlaul, and destroying some villages and crops in the Tarái, this division did nothing; and, without any exception, accomplished none of the objects for which it had been formed.*

* Yet Major-General J. S. Wood had belonged to the 8th Royal Irish Dragoons, a regiment second to none in glorious achievements. Evil communications will corrupt good manners, but the converse will not always create good sense.

[1] His own report. Papers, p. 525.

1815

Operations of the 1st Division, commanded by Major-General Bennet Marley.

4-22 R.A.

The artillery with this force was the 6th Company, 2nd Battalion.* The following officers were present:—

Major George Mason, commanding.

Captain	Alexander Lindsay.
Lieutenant	Wm. G. Walcott, Adjutant.
,, ...	Patrick G. Matheson.
,, ...	George Blake.
,, ...	Roderick Roberts.
Lieutenant-Fireworker	Richard R. Kempe.
,, ,, ...	George Twemlow.
,, ,, ...	William Counsell.

1814 November

See Plate XXXI.

If the result of the operations of the division under Major-General J. S. Wood was a failure, that of the 1st Division, to which a capital object of the campaign had been assigned, was still more abortive. In the month of November, Major, subsequently Lieutenant-Colonel, Paris Bradshaw, the Governor-General's political agent, was in command of the advance; and, while he directed Major Rougshedge, of the Rámgarh Battalion, to take possession of the Tarái of the Tirhut district, proceeded himself to occupy that of Sáran. On the 25th he attacked and defeated an outpost of the enemy at Barharwa, on the Bágmatti river; the Gurkha officer in command, Parsarám Thápá, being slain in single combat by Lieutenant J. Boileau, N.I. Lieutenant-Colonel Bradshaw then occupied three advanced positions close to the Bhábar—Báragarhi by a detachment under Captain C. P. Hay, of the Champáran Light N.I.; Parsa, 20 miles to the west, by another under Captain H. Sibley, 2—15th N.I.; and Samanpur, 14 miles to the east, by

* There was probably a detail attached from another company, as this one mustered 108 non-commissioned officers, rank and file—a little more than its regulation strength.

a third, under Captain J. F. Blackney, 2—22nd N.I. Each of these detachments was furnished with a 6-pounder gun, notwithstanding an order[1] prohibiting the practice of detaching single guns. Four 6-pounders, which had been drawn from two battalions of the division, were all singly disposed in this manner; the fourth being at Gorakhpur. Lieutenant P. G. Matheson was sent to command that at Parsa, and Lieutenant-Fireworker G. Twemlow at Báragarhi.

1814 November

Major-General Marley arrived with the main body of the division, and took up a position, on the 12th of December, at Pachrauta, in the rear of and about twenty miles from Báragarhi, but made no attempt to afford any support to their outposts, already too long exposed to attack. The rest of the month was spent in devising a plan for ascending the hills in three columns, but not in giving the enemy any reason to think he intended to fight. With strange infatuation, too, or prompted perhaps by a contempt for an enemy they had never met in fight, but whose courage had already been elsewhere displayed, Captains Sibley and Blackney omitted to secure themselves in their position. From such neglect the natural consequences ensued. Both posts were attacked on the morning of the 1st of January, and driven back with heavy loss: 258 casualties at Parsa, 125 at Samanpur. Captains Sibley and Blackney were both among the victims of their own rashness and the incapacity of the commander. There were no artillery at the last-named place, except the lascars who manned the drag-ropes. At Parsa, Captain Sibley, hearing of an intended attack, had applied for a reinforcement; but it arrived too late. His position was badly chosen, and exposed him to attack on all sides. The artil-

December

1815 January

[1] G. G. O. 8th November, 1806.

1815 January

lerymen, one by one, were all either killed or wounded; but Lieutenant Matheson fought his one gun as long as he had ammunition. Two names in his report cannot be passed over: in the present day they would have been recommended for a Victoria Cross. They were Matross William Levy, who, though wounded by a musket shot through one leg and one arm, yet gallantly continued to keep his station till the priming-pouch was blown from his side, and his wounds, becoming too painful to endure, obliged him to sit down; and Silári, a gun lascar of the 42nd Company, who, though wounded in both hand and foot, continued alone to assist Lieutenant Matheson to the last, and who seized and carried away with him a silver spear,* which the enemy planted close to the gun.[1]

The artillery casualties in this unfortunate business were:—

	European Artillerymen.				Gun Lascars.		Ordnance-drivers.	Puckallies.	Dooly-bearers.	Total.
	Sergeants.	Corporals.	Gunners.	Matrosses.	1st Tindals.	Lascars.				
Killed	1	...	2	1	...	3	3	...	...	10
Wounded	1	1	1	6	1	6	2	1	1	20
Missing	...	...	...	...	...	...	2	...	...	2
Total ...	2	1	3	7	1	9	7	1	1	32

The gallant conduct of Lieutenant Matheson was prominently noticed in the orders of the day. It was not the only instance in the course of this officer's service in which it was conspicuous.

* This spear was kept among the trophies in the Artillery Mess House, at Dum Dum. All these interesting relics have long ago disappeared.

[1] Lieutenant Matheson's report to Lieutenant C. Smith, January 6th, 1815. Papers, p. 531.

1815 January

After this, Major-General Marley gave up all idea of penetrating the hills. Though reinforced to a strength of 12,000 men, including a battering train of four 18-pounder guns, two 8-inch howitzers, and two 5½-inch mortars, which had arrived under Major George Mason,* and urged to action by the frequent letters of his commander-in-chief, he could come to no decision in his own mind but one, and that he carried out. Lord Moira had already determined to remove him from his command, but it was terminated more unexpectedly. Histories tell of armies running away from their generals, but there is, I believe, no record of a general running away from his army as General Marley did. Oppressed by a sense of responsibility which he could not bear, he left camp before daylight on the 10th of February without notifying his intention, or making over his office to any one.† Major-General George Wood arrived and assumed command upon the 21st, the day after a body of 500 Gurkhas had been attacked and cut up by a party under Lieutenant Pickersgill and Cornet J. B. Hearsey, at Pirári, close to camp. This intimidated the Gurkhas, who withdrew from the plains.

February

Dec.–March

There was another force, under the orders of Colonel R. B. Gregory, which was also under the command of General Marley, though not at first detailed

* This battering train left Cawnpore, under charge of Lieutenant-Fireworker George R. Scott, on the 23rd of November, and was joined at Fyzabad by Lieutenant J. Pereira, with some details of artillery from Allahabad.

† One cannot help contrasting General Marley's after treatment with that of the Gurkha chief, Bhagat Singh, who was recalled by his sovereign because he had not, with inferior forces, attacked the fortified post of Bárágarhi. General Marley lived in India drawing the high pay of his rank, and holding for many years the post of commandant of the Allahabad garrison. He died a full general at Barrackpore, June 14th, 1842. Bhagat Singh was obliged to attend the Rájá's durbar at Kátmándu in women's clothes.

1815 March C-19 R.A.

as part of the 1st Division. The 7th Company, 2nd Battalion was coming up from Dum Dum by river route, and Lieutenant Edward Huthwaite, doing duty with it, was detached to take charge of two 6-pounders which went to join Colonel Gregory. Lieutenant Thomas Marshall was also with this detachment. Its services had been called for by the civil magistrate of Tirhut, owing to the exaggerated reports of Gurkha incursions into that district. Colonel Gregory therefore marched over the northern portion of it, taking the posts of Rupitagarhi, Nárágarhi, and Bará Ruá. On his taking up his position at Náthpur, on the Kosi river, the two artillery officers were relieved by Lieutenant Roderick Roberts, and returned to Dinapore. The 1st Division did not advance into Nipál under Major-General George Wood, on account of the lateness of the season; and the treaty concluded with Nipál soon after temporarily suspended hostilities.

SECOND CAMPAIGN.

1816

Early in the following year, it became evident that the Nipál Government did not intend to abide by the conditions they had agreed to, and the Governor-General again assembled a force at Dinapore, which was placed under the command of Major-General Sir David Ochterlony, K.C.B. It was composed of four brigades, which were formed into three columns, as follows:—

CENTRE COLUMN, under the personal command of Sir David Ochterlony.

3rd Brigade—Lieut.-Col. F. Miller, 87th Regiment, commanding.
4th Brigade—Colonel G. Dick, 9th N.I.,* commanding.

* In the absence of Colonel Dick, who had not joined when the force advanced into Nipál, Lieut.-Colonel J. Burnet, 8th N.I., commanded this brigade.

1816

RIGHT COLUMN.

1st Brigade—Colonel W. Kelly, 24th Regiment, commanding.

LEFT COLUMN.

2nd Brigade—Lieut.-Col. C. Nicol, 66th Regiment, commanding.

The artillery detailed were as follows:—

5TH COMPANY, 2ND BATTALION. 4-23 R.A.

Capt.-Lieutenant	J. McDowell.
Lieutenant	J. E. Debrett.
Lieut.-Fireworker	G. R. Crawfurd.
"	G. Twemlow.

6TH COMPANY, 2ND BATTALION. 4-22 R.A.

Captain	A. Lindsay.
Lieutenant	George Blake.*
"	Roderick Roberts.†
Lieut.-Fireworker	R. R. Kempe.
"	W. Counsell.

7TH COMPANY, 2ND BATTALION. C-19 R.A.

Captain	John A. Biggs.	
Lieut.-Fireworker ...	W. Geddes.	
" ...	R. C. Dickson ‡	Doing duty from 2nd Company, 1st Battalion.
" ...	E. Huthwaite	

STAFF.

Lieutenant W. G. Walcott, adjutant.

Lieut.-Fireworker J. Cartwright, aide-de-camp and secretary to Sir D. Ochterlony.

January

The centre column, which was encamped in January at Lál Parsa, near Segowli, was to move direct upon the Nipalese capital, by Etounda; Colonel Kelly's, which was at Bhagwánpur, by Hariharpúr; and Colonel Nicol's, from Rámnagar, by the valley of the Rápti: the two latter to join the first at Etounda.

See PLATE XXXI.

To co-operate, a column was ordered to assemble at Sitapur, in Oudh, under the command of Colonel Jasper Nicolls, with orders to invade the province of Doti, east

* On command. † On command in Tirhut district.

‡ On command at Kissanganj, with Captain B. Latter's force.

1816 January

of the Káli, by way of diversion. The artillery detail for this service was large, but the column was not employed in any way, and need not be further noticed.

February

The centre column left Balwi on the 3rd of February, and halted for five days at Simora Básá, on the borders of the Bhábar, where a depôt was formed and the heavy guns were left. On the 9th it entered the forest, and next day took up a position at Bichiakoh. It appeared that the direct road into the hills from this place was occupied by the enemy; the next four days were, therefore, spent in reconnoitring. Lieutenant Pickersgill discovered another, though very difficult, path by which the pass might be turned. The 3rd Brigade accordingly marched at five on the evening of the 14th, leaving its tents standing, which, to conceal the movement, were at once occupied by the 4th, Colonel Burnet's brigade. The 3rd Brigade, entering the Bálikola ravine, followed its course five or six miles, then striking up a watercourse, came to a steep acclivity of three hundred feet in height, up which they clambered; the rear not reaching their ground, five miles beyond the crest of this pass, till 9 o'clock on the evening of the 15th. Till the 17th the pioneers were employed in rendering the path practicable for elephants without their guns, which were drawn up by hand. Sir David was now in rear of the enemy's position at Churia Ghát, and the movement facilitated at least the movements of Colonel Burnet's brigade, which marched on the 15th. The only opposition experienced was in reconnoitring the stockaded positions on the 16th, when Lieutenant and Adjutant W. G. Walcott was wounded severely. The last and most formidable position in the pass was evacuated during the night. The ascent of the ghát itself was a matter of great difficulty. The pioneers with great labour and per-

1816 February

severance constructed a long flight of steps, formed of trunks of trees laid transversely up the steep slope of the hill, on the 17th and 18th, to enable the elephants to proceed. One of them actually got up a part of this "ladder" with a gun upon his back, but the weight of it overbalancing him, he toppled over and was killed. The rest ascended without their loads, and without difficulty.*

On the 27th the 3rd Brigade moved from Etounda to a position in an open level ground immediately south of the hills covering the fortified heights of Makwánpúr, and was joined next day by the 4th Brigade, in time to assist in repulsing an attack made by the enemy in force. Our loss in this action was considerable; but there were no casualties among the artillery, though the fire of the guns told with considerable effect upon the enemy.

The right column, under Colonel W. Kelly, attacked on the 1st of March the Hariharpúr hill. Lieut.-Colonel J. O'Halloran was sent forward, and took possession of an important post at a few minutes before 6 o'clock, dislodging from thence a picquet of the enemy, who, in the words of Colonel Kelly—

"In very considerable force made a most desperate and obstinate attack to recover this point. I was therefore obliged to send a few companies to support the rear of the position which was

* This ladder, as General G. Twemlow terms it in a letter to me, has not been mentioned in any of the accounts that I have seen, except in the *Asiatic Journal and Monthly Register*, vol. ii. for 1816, p. 408; but the general informs me that it was a very creditable piece of work, in the skill with which it was constructed and the short time it took in the making. "In some places the heavy guns were dragged up by luff tackles fastened to posts wedged into the rocks on either side, with a preventer rope belayed on a third post in the centre, at the head of each ascent. Infantry soldiers manned the tackles, and a band of music played cheerily as they descended the slope; a subaltern of artillery sitting on the gun to direct the operation."—*Considerations on Tactics and Strategy by Colonel G. Twemlow*, 2nd Edition, p. 48.

1816 February

threatened. It was impossible, from the nature of the ground, to close or use the bayonet, and the musketry continued without interruption till half-past 11 o'clock, when the arrival of two 6-pounders and two 5½-inch howitzers on elephants in a few minutes decided the affair, and left us in possession of an almost natural redoubt. . . . . Amongst the wounded you will see Captain Lindsay, of the artillery. Although his wounds are not severe, I fear I shall lose his active services for a time,—which I lament exceedingly, having found Captain Lindsay a most zealous, able officer, both as an artillerist and an engineer."

Two lascars wounded completed the number of casualties in this action.

The left column, in the meanwhile, had experienced no opposition. It had reached the hills at Máhájogni on the 14th of February, and, proceeding westward for a short distance, crossed the Churighát range; and after taking possession of a strong stockade at Jogiyah which was abandoned on our approach, halted five days at Ekor, a little further on. The enemy had two posts further north, at Kádrang and Upádrang; so Major Lumley was left here with two battalions and Lieutenant Huthwaite's guns, while Brigadier Nicol proceeded on up the course of the Rápti to Makwánpúr, which he reached on the 1st of March, the day after the action there. A few days after this a Gurkha envoy presented to Sir David Ochterlony, on his bended knees, the ratified treaty of peace.

March

This was the first time that Bengal troops had been employed in hill warfare, and the lessons learned were valuable to those who could profit by them. At first the difficulties of ground and the hardihood of the enemy so wrought upon the inexperience of some and the fear of others that failure and disaster resulted. It was Sir David Ochterlony who showed how the scantily defended position occupied by the Gurkhas could be won, with comparatively small loss, by tactical management.

His caution and judgment were admirably displayed in both campaigns.

In no case do the artillery appear to have been a cause of unlooked-for delay, although much time was necessarily spent in getting the guns along. Infantry still more than in the plains must be the principal arm, but it seems hardly necessary to assert the necessity for the use of artillery in mountain fighting, wherever artificial or natural defences are to be forced. Yet we afterwards find an officer, who led one of the successful columns in this war, thus expressing himself when, as commander-in-chief, he excused himself for having sent a body of native infantry to force the Khyber pass without guns :—

"The great utility of artillery in the pass we have yet to learn, and, as an artillery officer, General Pollock will probably attach their full weight to that arm. To me it seems that a gun unlimbered and fired two or three times at an enemy perched on the side of an adjoining mountain, would run a great chance of being dismounted or overturned, or its carriage much shaken, by its own recoil; and that, whilst thus delayed, two or three men, and a horse or two, would be struck—thus adding to the delay of the column under fire."[1]

So difficult is it sometimes to recollect the lessons we have had to learn. In a note to this chapter (D) are given some details relative to ordnance prepared for this war, and its transport.

The services of the artillery may have been more brilliant in other campaigns, but there were none in which personal exertion and continued toil were more demanded from both officers and men of the fighting divisions than in the Nipál war.

[1] General Sir J. Nicolls to the Governor-General in Council, dated Loodiana, February 6th, 1842. No. 172, Papers relating to Military Operations in Afghánístán. Presented to both Houses of Parliament, by command of Her Majesty, 1843.

AUTHORITIES REFERRED TO IN THIS CHAPTER.

1. Papers respecting the Nipál War. Printed by order of the Court of Proprietors. 1 vol. folio.

2. Prinsep's History of the Political and Military Transactions in India during the Administration of the Marquis of Hastings. 2 vols. 8vo. London, 1825.

3. MS. Papers from Colonel C. P. Kennedy.

4. Minute by Colonel G. Brooke, C.B., on Mountain Train Batteries. From the Proceedings of Select Committee Artillery Officers.

5. Minute by Brigadier J. Tennant, C.B., dated Meean Meer, 9th June, 1852, on the same.

6. East India Military Calendar.

7. Copies of Muster Rolls.

8. Letters from Sir E. Huthwaite, Sir G. Brooke, Colonels Kennedy, Lumsden, Timbrell, and General Twemlow.

APPENDIX.

Note A.—Numerical strength of the divisions employed in the first campaign.

Note B.—General Gillespie's instructions to the storming parties at Kalanga.

Note C.—Strength of the artillery employed in the second campaign.

Note D.—Note upon the ordnance prepared for the war, and its transport.

Note E.—Muster-roll of the 2nd Troop Horse Artillery for October, 1814.

NOTE A.

Numerical strength of the divisions employed in the first campaign during the Nipál war, from Lord Moira's secret letter to the secret committee, dated 2nd August, 1815.

Divisions.	Commanders.	Cavalry.				Infantry.				Dromedary Corps.	Pioneer.	Artillery, European and natives, including Lascars and Drivers.	Ordnance.					
		European.	Native.	Irregular.	Total.	European.	Native.	Irregular raised for this Campaign.	Total.				Field Guns.	Field Howitzers.	Siege Guns.	Howitzers.	Mortars.	Total.
1st	Major-General Marley, afterwards General Wood *	...	...	...	...	3205	8802	...	12007	200	273	944	15	10	4	2	4	35
2nd	Major-General Gillespie, afterwards General Martindell †	...	...	100‡	100	856	8492	6668	16016	...	164	810	6	4	4	...	6	20
3rd	Major-General D. Ochterlony ...	...	...	...	...	...	5735	4463	10198	...	316	1061	10	4	2	...	6	22
4th	" J. S. Wood	...	564	...	564	...	3332	900	4232	...	90	712	8	2	2	...	3	15
	Captain Barré Latter	...	...	...	...	...	2622§	...	2622	...	...	101	...	Not stated.		—	—	—
	Col. Nicholls { Captain Hearsey ...	...	...	...	...	...	...	1080‖	1080	...	...	...	...			—	—	—
	Col. Nicholls { Lieut.-Col. Gardner	...	...	...	...	...	...	4000	4000	...	...	Strength not stated.	4	...	...	...	...	4
	Col. Nicholls { Col. Nicolls ...	...	...	...	...	...	2025	...	2025	...	...		4	...	2	...	4	10
	Total	...	564	100	664	4061	31,008	17,111	52,180	200	843	3628	47	20	14	2	23	106

* Strength after two reinforcements, including battering train.
† The (dismounted) detachment 8th Light Dragoons is unaccountably omitted.
‡ Of Skinner's Horse.
§ Including the Rangpur battalion, not a line regiment. The irregular infantry included in this column.
‖ The strength of Captain Hearsey's corps is taken from his letter to the adjutant-general, dated 18th Nov., 1814. Papers, p. 282.

Note B.

Extract of Field Army Orders by Major-General Sir Robert Gillespie, K.C.B.

"Camp near Deyrah, 29th October, 1814.

". . . . Officers will be careful to direct their men on all occasions to reserve their fire, and on no account to allow a shot to be fired at random; and the Major-General expects they will distinctly explain to their respective corps the necessity, in action, of taking a cool and deliberate aim, and above all to impress on their minds the advantage to be gained by a determined use of the bayonet.

"Officers at the heads of columns of attack will move deliberately, so that the men will not lengthen out, but be enabled to preserve their distance, and keep up without fatigue, or exhausting their breath. Officers therefore are recommended to bring their soldiers to the storm in possession of all their physical powers, to effect the impression that animal spirits and unimpaired vigour can always command. Strict silence to be observed; and if necessary to give a word of command during the march of a column to a point of attack, it must be communicated from the front to the rear by the men themselves repeating in a whisper the word of their commander.

"When the head of a column is prepared to debouch towards the point of attack, a short halt should be made to gain breath, if circumstances will admit, and the officers in command will bring up their men in compact order, with steady and cool determination.

"This is the moment an enemy will endeavour to take advantage of any looseness or precipitation. In all attacks against entrenched or stockaded posts (generally speaking) firing and halting to reload often cause severe loss. This may be avoided by an undaunted and spirited storm. In case of ambuscade, or surprise, a soldier requires all his natural courage, and when he is so situated as to be exposed to such attacks, in jungles and narrow pathways, he must predetermine within himself to preserve the utmost coolness; hurry must be avoided to prevent confusion; and even loss sustained with steadiness can be remedied. An officer in command ought always previously to arrange in what way he should repel and guard against such occurrences.

"The enemy we have to encounter are dexterous in using a short sword—Officers, caution your soldiers to keep them at the point of the bayonet; in the storm, beware of their closing.

"When several columns move to given parts, officers commanding them will bear in mind the utility and necessity of timing their march so as to render the attack simultaneous. The effect of several columns moving at once on an object is in most cases decisive.

"Let emulation actuate all; but corrected by steadiness and coolness, and no breaking of ranks or running for who is to be foremost in the contest. Each column must be a mutual support, and every soldier actuated by the principle of cool and deliberate valour will always have the advantage over wild and precipitate courage.

"Major-General Gillespie presumes to offer these suggestions notwithstanding the many excellent and experienced officers in the field might have precluded the necessity; he relies, however, on their indulgence, which he is confident he will experience, from the harmony and zealous, soldier-like feeling that appears to inspire all."

Note C.

The amount of artillery actually employed was less than the Marquis of Hastings, as commander-in-chief, in his letter of the 28th of November, 1815, had reported to Government as under orders for service. This was—

					Battering Train.					Light Field Train.			
	Field Officers.	Captains.	Subalterns and Staff.	N.C. Officers, Rank and file.	18-pr guns.	8-inch how.	10-inch mor.	8-inch do.	5½-inch do.	12-pr. guns.	6-pr. do.	5½-inch how.	4⅖-inch do.
With General Ochterlony:—													
European Artillery Golandáz ...	1	4	13	202 401	8	2	1	4	8	8	24	8	8
With Colonel J. Nicolls:—													
European Artillery Golandáz ...	1	1	7	104 59	4	...	...	...	4	2	6	2	2

The Light Field Train with General Ochterlony was calculated for eight infantry brigades; and there were in addition twelve 4⅗-inch howitzers and mortars for mountain service ordered.

The small mortars for Colonel Nicolls' force were exchanged for 24-pounder carronades, a species of ordnance which the Marquis of Hastings thought might be more generally utilized. Lieutenant Huthwaite was placed in command of a battery of them in the Pindári war, but they never came into action. They were mounted on 12-pounder field-gun carriages, adapted and strengthened for the purpose. The Marquis of Hastings, in his journal, says "that four bullocks could readily draw one, and that their range was nearly 2000 yards."

NOTE D.

Previous to the declaration of war with Nipál, the question of artillery equipment had engaged Lord Moira's attention. Some carriages for 3-pounder guns and 4-inch howitzers were made up, capable of being taken to pieces and carried separately. The trial made of them appears, from the journal of the Marquis of Hastings, to have been satisfactory. We find the following record in it:—

"*Cawnpore, October* 11*th*, 1814.—Colonel Grace had caught my conception so exactly, that I found the carriages of the howitzer precisely what I wished. The gun wheels can be taken from the body of the carriage within one minute, so as that the pieces are severally portable (slung on bamboos) by ordinary porters: the howitzer can be remounted with equal despatch. Its principal advantage is that it throws the shrapnel shell of a 12-pounder."

But neither the 3-pounder gun nor the howitzers or mortars of 4⅗-inch bore proved of much use. Want of penetration, shortness of range, and uncertainty of aim were all there, although the 3-pounder weighed three hundredweight. On this point I have before me now the testimony of Brigadier James Tennant and Colonel George (now Sir George) Brooke, in two minutes upon the report of a special committee, assembled at Umballah and Simla in 1851, to determine upon the best nature, equipment, and transport of ordnance intended for mountain service. Six-pounder field guns even were found of little avail at times; but the 12-pounder bronze gun of twelve or thirteen hundredweight did good service.* These pieces and their carriages were carried separately upon elephants. The carriages were heavier than those which came into use after 1836; yet, on Colonel Brooke's testimony, it appears that they did not stand, especially the field-pieces, the rough work they had to undergo.†

It has been seen that 18-pounders were sent with the 3rd Division, under Lieutenant Tennant. It will not be out of place to quote his notice of them from the same source, with reference to the *possibility* of employing them in mountain warfare. He says:—

"I may observe that, during Sir D. Ochterlony's campaign in the north-west, two 18-pounders were taken nearly 70 miles into the hills, over every description of acclivity that may be conceived to be practicable; and they were employed in the reduction of four stone castles, or forts, previous to his attacking Malown, which however, though fully as strong as any, was breached by two brass 12-pounders, and surrendered before the 18-pounders arrived. The progress of these pieces was necessarily slow, sometimes not exceeding one or two miles a day, for a road had to be prepared by the pioneers, and the guns to be dragged by hand, while all the ammunition had to be carried by men; and such was

* "I have myself fired for days from a 6-pounder of six hundredweight at a breastwork of loose stones, and at not more than 500 yards' distance, without apparently disturbing the enemy, who was, however, dislodged with half a dozen rounds the moment a 12-pounder (gun) arrived."—Brigadier Tennant in the minute referred to above. Colonel Brooke, from a similar experience, condemned the 12-pounder *howitzer*, which then had a charge only of eight ounces.

† "On reaching Malown . . . . nearly every one of the carriages, latterly, were merely held together by rope-lashing; when the piece became heated under firing we found the carriages upset every fourth round or so. This was not, as far as I can call to mind, attributed to the narrowness of the axletree, but more to a want of due proportion of height, and more to the whole machine being so light as to leap from the ground, as it were, and tumble over as a toy cannon generally does."—Colonel G. Brooke, ibid. This referred to the 12-pounder howitzers.

the expense and labour attending these operations, that when the army returned to cantonments, Sir David thought it better to leave the 18-pounders behind; and they remain in the Fort of Malown, near Simla, to the present day. Having been attached to these 18-pounders during nearly the whole of their progress, I can safely give my opinion that there is no country, however rugged, over which they might not be transported by like means. Time and expense are alone to be considered in effecting this object; but I hardly think that any occasion is likely to occur where we should again be justified in making such a sacrifice."

Both the above-mentioned officers appear to have been satisfied of the utility of elephants as transport animals, notwithstanding their behaviour at Sobraon, where they bolted. Their timidity is the great bar to their employment; it is liable to show itself not only under fire, but also in bad or difficult ground. They fear the report of small arms more than of guns—rockets and fireworks more than either; but cover can generally be found for them in hilly ground. "I have seen," Brigadier Tennant writes, "an elephant, with a gun loaded on his back, when ascending a narrow footpath up a steep rock to get the piece into battery, come to a dead stop in the middle of the ascent, roar and shake himself as if in great terror, while the mahout for a time lost all command over him. Still, I consider the elephant an invaluable animal for the carriage of artillery in hill service; and my experience makes me confident that he will travel with his load wherever man can find his way, provided such branches of trees as may obstruct his path are removed." Great care is necessary in the management of these animals, as they are very liable to suffer from sore feet or sore backs. Those with the park of Ochterlony's division, in 1814-15, were for some time exposed to snow without taking any harm. But little is known of the diseases and pathology of the elephant: from time immemorial they have been treated by the head mahout according to an unvarying pharmacopeia, dating certainly from the time of Akbar Shah, if not of Porus; and, as these men jealously guard their written recipes from impertinent criticism, I venture to suggest the subject as one deserving our attention.

NOTE E.

Muster-roll of the 2nd Troop Horse Artillery, commanded by Captain Harry Stark, for the month of October, 1814.

Meerut, 1st November, 1814.

	Rank and Name.	Remarks.
	CAPTAIN.	
1	Harry Stark	On command. Marched 4th Oct., 1814.
	CAPTAIN-LIEUTENANTS.	
1	John Peter Boileau	On command at Java.
2	William Samson Whish	On command. Marched 12th Oct. Returned 20th Oct., 1814. Commanding the station of Meerut.
	LIEUTENANTS.	
1	Hugh Lyon Playfair	Adjutant and quarter-master. On leave of absence. Returned on the 31st ultimo. Pay not drawn.*
2	John Curtis	On command. Marched 4th ultimo.
3	Thomas Lumsden	On command. Marched 14th, returned 16th ultimo.
4	John Sconce	Appointed to the 3rd Troop. On command. Marched 14th ultimo.

Muster-roll of a detachment from the 2nd Troop of Horse Artillery, commanded by Captain H. Stark, for the month of October, 1814.

Cawnpore, 1st November, 1814.

	Rank and Name.	
	CAPTAIN.	
1	H. Stark	
	LIEUTENANTS.	
1	John Curtis	
2	James C. Hyde	

* Lieut. Playfair returned from leave to Calcutta, and joined the 3rd Troop before Kalanga 3rd November. He was slightly wounded on the 25th *idem*, in the 18-pounder battery, but was not returned in any casualty roll.

NOTE F.

Names of officers of Bengal Artillery employed in the first campaign against Nipál in the year 1814-15.

HORSE ARTILLERY.

Served with the 2nd Division in the advance on Dera and assault of Kalanga. Only four guns of each troop proceeded on this service.

Rank.	Names.	No. of Troop.	Remarks.
Captain (Brevet-Major)	Gervaise Pennington	1	Commanding Horse Artillery
Captain (Brevet-Major)	James H. Brooke	3	
Capt.-Lieut.	John Rodber	3	
Capt.-Lieut.	William McQuhae	1	Adjutant and Quarter-Master of Reserve
Lieutenant	H. L. Playfair	2	
"	Charles P. Kennedy	1	
"	Gabriel N. C. Campbell	1	
"	John B. B. Luxford	3	
"	R. S. B. Morland	3	

FOOT ARTILLERY.

Rank.	Names.	Co.	Batt.	Remarks.
Major	George Mason	—	2	Commanding Artillery 1st division
"	Alexander Macleod	—	3	Commanding Artillery 3rd division
Captain	George Pollock	3	3	4th division
"	Alexander Lindsay	6	2	1st division
Capt.-Lieut.	William Battine	6	3	Commanding Artillery 2nd division after Major Pennington left
"	John McDowell	5	2	Commanding Artillery 4th division
"	Nathaniel S. Webb	4	3	3rd division
Lieutenant	James Tennant	6	3	3rd division. Joined after capture of Nálágarh
"	Isaac Pereira	3	2	1st division
"	Charles Graham	4	1	3rd division. Joined after capture of Nálágarh
"	Theodore Lyons	6	3	2nd division
"	Charles H. Bell	4	1	Colonel Nicolls' force in Kamáon
"	Thomas Marshall	3	3	Colonel Gregory's detachment in Tirhut

NOTE F (*continued*).

Rank.	Names.	Co.	Batt.	Remarks.
Lieutenant	William G. Walcott	2	3	1st division
"	John E. Debrett	5	2	Captain Barré Latter's force
"	Edward Hall	2	3	2nd division
"	George Brooke	4	3	3rd division
"	Thomas Croxton	2	2	4th division. Doing duty with 5th Company, 2nd Battalion
"	Patrick G. Matheson	3	2	1st division
"	Thomas Timbrell	7	3	3rd division. Joined after capture of Nálágarh
"	Charles C. Chesney	6	3	Adjutant and Quarter-Master 2nd division
"	George Blake	6	2	1st division
"	Roderick Roberts	6	2	1st division; afterwards Colonel Gregory's detachment in Tirhut
"	Kender Mason	4	3	3rd div. Absent sick (?)
"	Kenneth Cruikshank	4	3	Adjutant and Quarter-Master 3rd division
"	Charles Smith	5	3	2nd division
Lieut. Fireworker	Edward P. Gowan	7	3	2nd division
"	John Cartwright	4	3	3rd division
"	Rowland C. Dickson	7	2	With Captain Barré Latter's force in the second campaign
"	Edward Huthwaite	2	1	Colonel Gregory's detachment in Tirhut
"	Gavin R. Crawfurd	5	2	4th division
"	Robert B. Wilson	2	2	Colonel Nicolls' force in Kamáon
"	Tuneus A. Vanrenen	2	2	1st division
"	Richard R. Kempe	6	2	1st division
"	George Twemlow	5	2	1st division
"	Charles G. Dixon	5	3	2nd division
"	William E. J. Counsell	6	2	1st division

CHAPTER XI.

SIEGE OF HÁTHRAS, 1817—Dyarám and Bhagwant Singh—Háthras invested—Constitution of the force—Attack upon the town—Batteries—Town evacuated—Attack upon the fort—Batteries—Parallel opened—Bombardment commenced—Explosion of a magazine—Escape of Dyarám—Surrender of neighbouring forts—Orders published—Sir John Horsford's character and death.

SIEGE OF HÁTHRAS.

1817 DYARÁM and Bhagwant Singh, two zamindárs of the Doáb lying between the Jumna and the Ganges, had been allowed, when this portion of the North-West Provinces was made over to us by Sindiah, to retain possession of Mursán and Háthras, both strong forts, and relying upon the security they afforded, had for some years pursued a course of oppression and robbery, their defiance of law increasing with time and the presumed forbearance of Government. Dyarám, by copying every improvement of modern fortification introduced into the works of the neighbouring fort of Aligarh, thought to acquire for Háthras the reputation still enjoyed by Bhurtpore, of being impregnable.

But Lord Hastings, unwilling to leave two such rebels in the rear of his intended operations against the Pindáras, gave orders for the concentration of a strong force to be employed against them, from the Cawnpore, Muttra, and Meerut divisions of the army. No Governor-

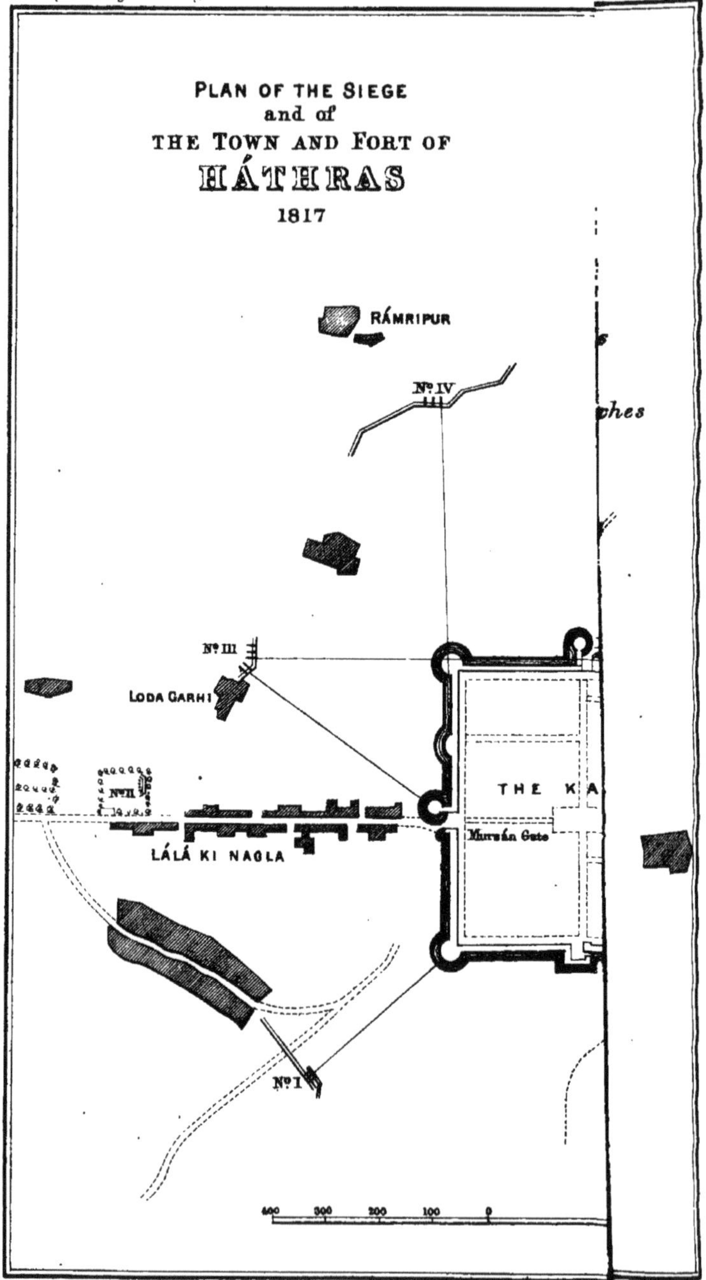

PLAN OF THE SIEGE
and of
THE TOWN AND FORT OF
HÁTHRAS
1817

RÁMRIPUR

Nº IV

Nº III

LODA GARHI

Nº II

THE K A

Mursán Gate

LÁLÁ KI NAGLA

Nº I

400 300 200 100 0

London: Henry S. King & Cº, 65, Cornhill

Edwᵈ Weller, Litho

General ever acted more fully upon the principle that the best economy is to complete a work thoroughly. He laid it down as an indisputable fact, that former failures in sieges had arisen from want of means, and determined that this should not be the case before Háthras. The instructions given in the artillery and engineer departments were therefore in accordance with these views.

1817

On the 12th of February, Háthras was invested by the three divisions: Major-General Dyson Marshall's, from Cawnpore, encamping on the east; Major-General Rufane Donkin's, from Muttra, on the south-west; and the Meerut force, under Major-General Thomas Brown, on the south. The first-named officer was in chief command. During the four days' negotiations which preceded actual hostilities, the engineers were allowed to reconnoitre, except between the *katra* (town) and fort, the officers of the army being allowed even to walk up the glacis and look down into the ditch; possibly with the humane intention of letting them change their minds before attempting to crack so hard a nut. The mean width of the ditch was about a hundred feet, and its depth from forty to seventy feet. The escarp was everywhere revêted with masonry, an embankment being thrown up at its foot, to protect the foundations. The nature of the defences will be understood better from the plan (Plate No. XXXIV.) than from a detailed description.

February

February 18th.—The force was brigaded this day as follows:—

ARTILLERY.

Major-General Sir John Horsford, K.C.B., commanding.
Captain Charles Hay Campbell, Brigade-Major.

1817 February

HORSE ARTILLERY.

Major Gervaise Pennington, commanding.
Lieutenant T. Lumsden, Adjutant and Quarter-Master.

A-C,R.H.A.	1st Troop*	Captain J. P. Boileau.
B-C,R.H.A.	3rd Troop*	„ J. H. Brooke.
B-F,R.H.A.	Rocket Troop	„ W. S. Whish.

FOOT ARTILLERY.

Major George Mason, commanding.
Lieutenant-Fireworker H. J. Wood, Adjutant and Quarter-Master.

Reduced in 1825	2nd Company,	2nd Battalion	Capt.-Lieutenant Alex. Fraser.
Reduced in 1825	3rd „	„ „	„ William Curphey.
Reduced in 1825	4th „	„ „	„ Edward Pryce.
4-22 R.A.	6th „	„ „	Captain Alex. Lindsay.
8-22 R.A.	4th „	3rd Battalion	Lieutenant Charles H. Bell.
B-19 R.A.	6th „	„ „	Capt.-Lieutenant William Battine.
Reduced in 1871	7th „	„ „	„ Wm. Tallemach.

1st, 2nd, 3rd, and 4th Companies of Golandáz.
18 companies of Gun Lascars.

ENGINEERS.

Major Thomas Anbury, commanding.

CAVALRY.

Major-General T. Brown, commanding.

1ST DIVISION (Colonel Newbery, 24th Light Dragoons, commanding).—8th Royal Irish Dragoons; 3rd Native Cavalry, and 1st Rohilla Horse.

2ND DIVISION (Lieutenant-Colonel Philpot, 24th Light Dragoons, commanding).—24th Light Dragoons; 7th Native Cavalry, and 2nd Rohilla Horse.

---

* Owing to the paucity of men with these troops, it was found necessary to transfer, as a temporary measure, half the men of the 2nd, or Captain Stark's Troop, for duty to the other two. The same thing was done when the 1st Troop went to Alwar in 1813, and afterwards in 1825, when a force was being collected for Bhurtpore.

1817
February

INFANTRY.

Major-General Rufane Donkin, commanding.

1ST DIVISION (Colonel James Watson, 14th Regiment, commanding).—14th Regiment; 2nd Grenadier Battalion N.I., and 2—11th N.I.

2ND DIVISION (Lieutenant-Colonel Vanrenen, 12th N.I., commanding).—87th Regiment; 2—12th and 2—15th N.I.

3RD DIVISION (Lieutenant-Colonel Cooper, 1st N.I., commanding).—1—1st, 2—25th, and 2—29th N.I.

February 19*th*, 20*th*.—The siege train from Cawnpore arrived, under Major-General Sir J. Horsford, with the engineers' park, escorted by the 14th and 87th Regiments and 2—15th N.I. The siege ordnance consisted of—

24-pounder guns	6	
18-pounder „	14	
	—	20
8-inch howitzers	4	
	—	4
10-inch mortars	6	
8-inch „	14	
5½-inch „	22	
	—	42

There were several field-pieces besides; and the twelve horse artillery guns were increased by the six gallopers of the 24th Dragoons,* the 3rd and 7th Native Cavalry, which were attached. The rocket troop brought 650 of its own projectiles.

February 21st.—Unpacking stores as they arrived. Heavy rain in the evening. It being determined to attack the *katra* first, four batteries were commenced upon, viz.:—

* The 8th Royal Irish Dragoons had not any gallopers in India. They refused to receive them, stating that guns would only impede their action as cavalry.

1817 February

No. I.—Three 24-pounder guns, about 300 yards from the south-west bastion.

No. II.—Four 5½-inch mortars, 600 yards from the centre bastion on the west.

No. III.—Three 18-pounder guns, north of the village of Lodagarhi, 330 yards from the north-west bastion.

No. IV.—Three 18-pounder guns in front of Rámripur, 450 yards from the north-west bastion.

February 22nd.—The guns were moved into battery by sunrise, and platforms laid down by the gunners. The mortars were not sent, as the fuzes were found to have been injured by the damp, and required to be dried and re-primed. In Nos. I. and III. batteries, commanded by Captains Lindsay and Battine, the embrasures were faulty in dimension or direction. In the former the gunners were delayed till 11 a.m. remedying this; and in the latter an ineffective fire was kept up till the evening, when the engineer on duty took down and rebuilt, during the night, the parapet from the sole of the embrasures. The battery was also lengthened towards the village of Lodagarhi for two 24-pounders in addition, in consequence of a gun on the gateway bastion having become troublesome. A gunner was mortally wounded in this battery to-day. Captain Gowan was sent by Major Pennington, with one 12-pounder and two 6-pounders, to a village on the right of No. I., to keep down the enemy's fire with shrapnel. No. IV. was exposed to the enfilade fire of a 32-pounder from the fort, which commenced early, and kept up a smart fire; so Captain Fraser, while waiting the signal to open fire from Captain Lindsay at No. I., caused his men to throw up an *epaulement*, which saved some lives.

Immediately on the right of No. III., Captain Whish took up a position with his rocket battery, and opened at 10.30 a.m. with carcass and shell rockets (32-pounders) from three frames. Of 138 discharged this day, about one-fourth took effect in the *katra*, setting it on fire in several places, and one-eighth in the fort; the remainder going either wide or too high. It was remarked, however, that the effect would have been still greater, had there been the means of discharging the rockets in larger numbers. As it was, they were found to have contributed in some degree to success. The very wildness of their flight was an additional source of terror.

1817 February

February 23rd.—Major Butler field-officer on duty. No. I.—Captain Tallemach continued breaching. No. II. —Lieutenant Croxton was not in battery with the mortars till after sunrise, and opened fire at half an hour after noon. After sunset fired a round every five or six minutes, when orders were given to keep up a rapid fire. No. III.—Captain Curphey, on relieving Captain Battine, was obliged to take up all the platforms of the 18-pounders and oblique them to the left, to coincide with the direction of the new embrasures; and when day broke, it was found that the 24-pounders could not be used with useful effect from the faulty construction of the battery, and they therefore remained silent. A 6-pounder gun was brought up to check the enemy's fire. Firing was kept up during the night every quarter of an hour, one round of grape to three of solid shot. Lieutenant Hele, in No. IV., assisted No. III. in breaching the N.W. bastion. The vents of two guns were injured, and the fire was slackened about 2 p.m. A lascar and a pioneer were wounded. Captain Whish only fired twenty rockets this day.*

* Artillery Brigade after-orders of this day :—" Each officer commanding a battery on being relieved will deliver or send in a written journal of all occurrences during the time he was on duty, to the acting Major of Brigade, for the Major-General's information.—(Signed) C. H. Campbell, Acting Brigade-Major Artillery."

1817 February

February 24th.—At 3 a.m. Lieutenant R. Tickell (engineers), under a fire of blank cartridge, examined the breaches and ditch, plumbing the latter with a 6-pounder shot. It was 24 feet deep. A storming party had been ordered to parade the evening before, but the report did not warrant the attempt. The enemy, however, evacuated the place early this morning, and it was at once occupied by our troops. The fire of the mortars was considered to have made a good impression, as the enemy had still ample means for resistance when they abandoned the place.

The project for the attack of the fort itself was to run a parallel from the south-east corner of the *katra* to the salient of Gopálghar. This work was connected with the main body of the place by two weak curtains, in one of which was a large opening. It was intended, after having blown in the counterscarp, and having effected a lodgment in Gopálghar, to throw up a second set of breaching batteries within, where they would be comparatively secure from the fire of the enemy. An attack upon the western side of the fort was to cover the design.

February 25th.—During the night the following batteries were marked out:—

No. V.,* for three 24-pounder guns, on the N.E. bastion of the *katra.*

No. VI., three 24-pounder guns, on the east rampart.

No. VII., one 18-pounder gun, on the right of the gateway, same face.

No. VIII., six 10-inch mortars, on the left of the road leading out of this gate.

No. IX., eight 8-inch mortars, also in advance, and between Nos. V. and VI.

---

* These numbers are given in reference to the plan in this work, but in the official records of the siege the numbering for the second attack commenced afresh. Thus Nos. V. to IX. were called Nos. 1 to 5. The rocket battery and that armed with horse artillery guns had no numbers, so XI., XII., XIII., and XV. were 6, 7, 8, and 9.

1817 February

The night was chiefly employed in clearing the rampart of the houses which blocked it up, and opening embrasures in the wall; the mortar batteries were also well advanced. Captain Tallemach, commanding the gun batteries this day, dragged the ordnance for Nos. V. and VI. up one by one, as the road through the town was very bad, and was also enfiladed from the fort. They opened at 4 and 5 p.m., and ceased firing at dusk, when the wall was so much damaged that the gunners in the former battery were quite, and in the latter were very much, exposed.

February 26th.—The gun batteries were repaired during the night with fascine work. Captain Curphey, commanding, reported that his fire had little effect against the enemy's guns, owing to their being so well concealed. Lieutenant Timbrell opened with the 18-pounder in No. VII. at 10 a.m. Captain Battine, on taking over No. VIII., laid his platforms at daybreak, and opened fire about 11 a.m. Captain Lindsay, in No. IX., found the platforms laid, but was delayed in getting the mortars into the battery, having the magazine enlarged, and the *epaulement* of the battery lengthened. Opened fire at 8 a.m. Captain Whish selected a position for a rocket battery (No. X.) about 100 yards in advance of the *katra*, and on the left of the road leading from the Sásni gate. It was begun at noon, and finished in five hours. It was arranged to contain 16 rockets, laid for firing at an angle of 28 degrees. Eleven were fired at sunset, to try their effect; the greater number entered the fort and set fire to some buildings, which burned for nearly an hour after. The plunging fire from the fort into the town was annoying, and caused some casualties. Captain Carmichael Smyth and Ensign Hutchinson, of the engineers, while reconnoitring without an escort in

1817 February

the afternoon, were nearly cut off by a party of the enemy.*

February 27th.—About 300 yards of the parallel were opened during the night. The working party consisted of 400 Europeans, 400 natives, and 100 pioneers, with a covering body of 400 men. The working party during the day was 100 men, who carried the parallel on for 150 yards more, towards some broken and excavated ground south of the fort. No. XI., for six 8-inch and six 5½-inch mortars, to the left of the excavations, was completed before daylight, and the broken ground itself was levelled for sixteen 5½-inch mortars (No. XII.).

Except the working party, the besiegers were at rest; and at midnight, when all was quiet within the fort, Captain Whish fired a salvo of 24 rockets. The wind was unfavourable; three or four took a northerly direction clear of the fort, some went over it, the rest inside. It had the effect of waking up the garrison, who immediately commenced a furious fire of guns, jingals, and matchlocks, and kept it up for an hour. The result was the expenditure of a good deal of very inferior powder.

The mortars were got into No. XI. by Captain Tallemach, who fired a few rounds into the *fausse-braie* and interior of the fort, to ascertain his range. Nos. VIII. and IX. were silent.

February 28th.—Working party the same as the previous night, with double the number of pioneers.

* Artillery Brigade Orders, 26th February: "The horse artillery having volunteered their services in the batteries, through Major Pennington commanding, the six 10-inch mortar battery will be manned by them to-morrow; and Major Mason will intimate to Major Pennington whenever he may be desirous to avail himself of the services of the officers and men.—(Signed) C. H. CAMPBELL, Acting Brigade-Major Artillery." Although "the mounted branch" has always shown the same zeal and readiness to serve out of their own line, the history of the siege would be incomplete without this record.

1817 February

The parallel was widened and improved, and No. XIII., for six 18-pounder guns, was marked out on the edge, and to the right of the excavation. This was intended to take in reverse and enfilade the defences of Gopálghar and its *fausse-braie*. Captain Smyth, with 100 pioneers, was sent to erect a battery (No. XIV.) for the horse artillery guns (four 12-pounders, four 6-pounders, and four 5½-inch howitzers),* at about 1000 yards from the northern face of the fort. It was seen that the enemy took refuge in the houses outside the Medhu gate on that front, which also gave shelter to matchlock-men. The enemy kept up a heavy fire during the night upon Captain Smith's party, and during the day upon the gunners as they laid the platforms, but to no purpose. Five companies of the 12th N.I., occupying Ganári, were kept there in support.

March

March 1st.—A second *banquette* was made behind the gabion parapet, skirting the excavations for musketry fire. The parallel was carried 50 yards further to the right, and No. XV., for six 24-pounders, was marked out to the left of No. XI. The working party consisted of 200 Europeans and 300 natives during the night, reduced during the day to 100 sepoys, with three companies of pioneers. Captain Pryce, commanding Nos. V. and VI., after firing 25 rounds, received orders to withdraw the ordnance to the engineers' depôt south of the town, for the purpose of arming another battery. Platforms for No. XII. were laid, and a few rounds fired in the evening to test the range. The whole of the mortar batteries were silent as yet, as it was intended, when all were ready, to commence a general bombardment. Captain Fraser, in No. XIII., dismounted one

* Batteries of horse artillery then consisted of two light 12-pounder guns, two 6-pounder guns, and two 5½-inch howitzers.

1817 March

of the enemy's guns, and silenced two others which had been troublesome. The horse artillery left camp at day-break and got into their battery at sunrise. The fort opened a desultory fire upon them, but as no notice was taken of it, it ceased during the day. A sufficient number of fuzes having been now prepared,* it was intimated that the bombardment would commence next day.

March 1st, 2nd.—The working party this night consisted of 200 Europeans and 300 sepoys, to which at 8 p.m. a company of pioneers, and at 11 p.m. 50 more sepoys, were added. The parallel was pushed forward nearly 300 yards, the mortar batteries were improved, and No. XV. was completed during the night. The different batteries to-day were commanded as follows:—

No. VIII.	Captain Lindsay.
,, IX.	,, Battine.
,, X.	,, Whish.
,, XI.	,, Fraser.
,, XII.	,, Tallemach.
,, XIII.	,, Pryce.
,, XIV.	Major Pennington and Capt. Gowan.
,, XV.	Captain Curphey.

At 9 a.m. a salvo from the rocket battery gave the appointed signal, and the whole burst into action. The fire of forty-two mortars and twenty-four guns and howitzers concentrated upon so limited a space can only be conceived by those who have witnessed such a scene. In scarcely ten minutes flames broke out in several different parts of the fort; the fire of the enemy, at first active and well directed, soon sunk, as the garrison sought for shelter from the deadly

* From the returns it appears that the lascars in the park drove the following number of fuzes between the 22nd of February and the 1st of March:—590 10-inch, 1408 8-inch, and 2430 5½-inch mortar. Those brought from Agra had been spoiled by damp.

shower. The conflagration within increased, and loose horses running on the parapet added to the fright and confusion. Several explosions took place, and at five o'clock one of the principal powder magazines blew up. The percussion shook the camp, and was felt, it was afterwards reported, as far as Agra, Delhi, and Meerut. A dense column of red smoke, said an eye-witness, shot up into the air, spreading out like an umbrella, in a rolling volume, from which a shower of stones fell, ranging as far as the trenches. The fire of the batteries was momentarily suspended, but was soon resumed, while that of the fort broke out into active but temporary life. The bombardment was continued after dark till nearly midnight, when orders to cease firing came round, and the drums beating within the fort showed our people were in possession. The Rájá, with forty or fifty followers, had got out and successfully run the blockade. Within the fort the scene of destruction and mortality defied description. Not a building had escaped. The crater formed by the explosion was eighty-three feet in its shortest diameter. Strangely enough, another powder magazine in the *fausse-braie* was found perforated by a shell, but the concussion had shaken out the fuze. One of its walls formed part of the escarp. The charred and disfigured remains of some 400 men and 80 horses, with 700 prisoners, remained to represent the garrison which that morning, though reduced by numerous desertions, had been more than 1400 men. On our side the casualties were from the commencement only seven killed and ten wounded (Appendix, Note C).

The fall of Háthras was followed by the immediate surrender of the neighbouring fort of Mursán, with eleven other smaller places near. The manner in which the

1817 March

operations were planned and carried out reflected the highest credit upon the engineer and artillery corps. The share of duty which fell to the rest of the force was not light, especially on the infantry, as in addition to the large working parties, a covering one, increased as the trenches extended from 400 to 1200 men, was constantly on duty. All the details appear to have been carefully carried out, leaving as little opening as possible to failure or loss.

The following extracts of orders referred to the artillery:—

"FIELD ARMY ORDERS.

"7th March.

"The science and skill displayed by the engineer and artillery departments were eminently conspicuous; and the bombardment and explosion of the enemy's principal magazine, which, without deteriorating from the merits of others, must be allowed to have given us almost immediate possession of the place, will be regarded as the most memorable events of the last fortnight, and as demonstrative of the extent and soundness of that judgment and penetration which, in the avowed anticipation of these very consequences, enabled the army, by the provision of adequate means, to ensure them. The practice of the artillery has answered the expectations of that high authority to which the Major-General has ventured to allude in the foregoing observations."

"ARTILLERY BRIGADE ORDERS.

"6th March.

"Major-General Sir John Horsford, commanding the artillery brigaded before Hatras, performs with pleasure the last exercise of his command, in conveying to the horse and foot artillery and rocket troop his congratulations on the brilliant services which their united exertions have effected for the State—Government having, through their means principally, been placed in possession of Hatras, the rapid reduction of which has caused the surrender of the important fortress of Mursan and eleven other forts.

"The acknowledgements of Major-General Marshall, commanding the army, and the favourable sentiments entertained by the army at large, must be more satisfactory to the artillery, than any tribute of praise which Sir John Horsford could bestow in confirmation of their meritorious services.

1817 March

"But the Major-General considers public acknowledgements due to Major Mason, commanding the foot artillery, who with Majors Macleod and Butler superintended in turn the several batteries. He begs to offer his thanks to Major Mason, and to the experienced field officers above mentioned, for their several important services.

The Major-General duly appreciates the labour and exertion of every officer and man employed in the batteries before the *kutra* and fort, and more particularly, the heavy duty all had to perform on the 2nd instant, during the general bombardment. To the officers commanding batteries, and to their juniors doing duty under them, the Major-General's notice is particularly due. The state of the fort, after its capture, evinced to all that the means employed for its reduction had been directed by hands well acquainted with their use. When every officer was equally zealous, the Major-General hopes he will be excused for not naming all who deserve his public thanks.

"The mature judgment of Major Pennington was displayed on every occasion which offered itself.

"The spirit and conduct of the officers and men of the horse artillery throughout the service deserve the Major-General's warmest approbation."

Same date.

"The zeal evinced by Captain Whish, the officers and men of the rocket troop, requires the Major-General's notice in public orders.

"The Major-General's personal feelings are much gratified by the important consequences which have resulted from the unanimity which prevailed amongst every branch of the artillery to forward the objects of the service. The preparations which were made caused much to be expected from their exertions, and the Major-General is satisfied that the expectations of the most sanguine have been realized. It is a source of great pleasure for the Major-General to reflect, that the last period of his service with a corps in which he has long served should be distinguished by events which call forth the admiration of all who witnessed them, and by services which conspicuously increase the credit and the established high character of the Regiment of Bengal Artillery.

(Signed) "C. H. CAMPBELL,
"Acting Brigade-Major Artillery."

The siege of Háthras was a fitting termination to the military services of the distinguished officer who then commanded the Bengal Artillery. It was the last period

1817 March

of his service with the corps, though not from the cause that was anticipated. Forty-four years had passed since John Horsford came out as John Rover, artillery recruit, in the *Duke of Grafton* East Indiaman, to avoid the profession which had been chosen for him. His character, for the six years during which he served in the ranks, showed that the circumstances of his enlistment had produced no bad fruit. From the year 1778, in which he was first commissioned fireworker, till his death in 1817, he was noted throughout the army for his extensive acquaintance with all the details of his profession, his habits of system and application, and his perfect integrity. Remarkably abstemious and temperate, he never during his service had taken leave of absence even for a day, and, like the soldier who commanded the regiment when he joined it, he devoted himself to improving the character and efficiency of the corps. To Colonel Pearse and Sir John Horsford must be assigned the foremost place among those officers who gained its distinguished name for the Bengal Artillery. Sir John Horsford died suddenly at Cawnpore, on the 20th of April in this year, of ossification of the heart.

In the old burial-ground of that station, which fortunately escaped the destroying hands of the rebels in 1857, are two large monuments. One, a pillar, stands near the western end; the other, a sarcophagus underneath a dome, near the entrance at the eastern end. They are those of Sir Dyson Marshall and Sir John Horsford, the two senior officers of the force that captured Háthras.

AUTHORITIES CONSULTED FOR THIS CHAPTER.

1. Journal of the Siege. Published in the *East India United Service Journal* for 1837.
2. East India Military Calendar.
3. Copies of Muster-Rolls.

APPENDIX.

NOTE A.—Officers of the Bengal Artillery who served at the siege of Háthras.

NOTE B.—Expenditure of ammunition from heavy guns.

NOTE C.—Casualties during the operations.

NOTE A.

Officers of the Bengal Artillery who served at the siege of Háthras.

Major-General Sir John Horsford, K.C.B., commanding.
Captain Charles Hay Campbell, Aide-de-camp and Brigade-Major.

HORSE ARTILLERY.

Major Gervaise Pennington, commanding.
Lieutenant Thomas Lumsden (1st Troop), Adjutant.

1ST TROOP.	3RD TROOP.
Captain John P. Boileau.	Captain James H. Brooke.
Captain-Lieutenant G. E. Gowan.	Captain-Lieutenant John Rodber.
Lieut.-Fireworker R. S. B. Morland.	Lieutenant James C. Hyde.
„ G. Pennington.	„ Donald Macalister.
	„ John Sconce.

ROCKET TROOP.

Captain William S. Whish.
Lieutenant George Brooke.
Lieutenant-Fireworker John Cartwright.

FOOT ARTILLERY.

Major George Mason.
„ Alexander Macleod.
„ Edward W. Butler.
Lieutenant-Fireworker Henry J. Wood (7th Company, 3rd Battalion).
„ Charles G. Dixon (Adjutant, Agra Division of Artillery).

2ND COMPANY, 2ND BATTALION.	4TH COMPANY, 2ND BATTALION.
Capt.-Lieut. Alexander Fraser.	Capt.-Lieut. Edward Pryce.
Lieut. Thomas Croxton.	Lieut. Thomas Chadwick.
Lieut.-Fireworker G. R. Scott.	„ John C. Carne.
„ R. B. Wilson.	Lieut.-Fireworker T. Sanders.
	„ J. D. Crommelin.

3RD COMPANY, 2ND BATTALION.
Capt.-Lieut. William Curphey.
Lieutenant Isaac Pereira.
Lieut.-Fireworker J. S. Hele.
„ Tuneus A. Vanrenen.

6TH COMPANY, 2ND BATTALION.
Captain Alexander Lindsay.
Lieut. Samuel Coulthard.
„ Roderick Roberts.

4TH COMPANY, 3RD BATTALION.
Lieutenant Charles H. Bell.
„ Charles Smith (doing duty from 6th Company, 1st Battalion).
Lieutenant-Fireworker Lucas Lawrence (doing duty from 1st Company, 2nd Battalion).
„ Charles R. Whinfield (doing duty from 7th Company, 1st Battalion).

6TH COMPANY, 3RD BATTALION.
Capt.-Lieut. William Battine.
Lieut. T. Dingwall Fordyce.
Lieut.-Fireworker R. C. Dickson.
„ H. Delafosse.

7TH COMPANY, 3RD BATTALION.
Captain William Tallemach.
Lieut. Thomas Timbrell.
Lieut.-Fireworker E. P. Gowan.

NOTE B.

Expenditure of ammunition, during the siege of Háthras, from heavy guns and mortars.

Shot.—24-pounder gun	1745
18 „	2730
Shell.—10-inch mortar	468
8 „	930
5½ „	1711

NOTE C.

Casualties during the operations before Háthras.

IN THE ATTACK UPON THE TOWN.

Killed.—One golandáz.

Wounded.—One gunner, one pioneer, and one lascar; all mortally.

IN THE ATTACK UPON THE FORT.

Killed.—Two dragoons; one náik of native infantry.

Wounded.—Four dragoons; one gunner; two sepoys; one lascar; one ordnance driver, and one bildar.

CHAPTER XII.

PINDÁRI AND MÁHRÁTÁ WAR, 1817-1819—Assembly of the army—Its constitution—Positions of the Divisions at the beginning of the war—Detail of the Centre Division grand army—Of the Right Division—Of the Left Division—Of the Reserve—Of Generals Hardyman and Toone's columns—Of the 5th Division army of the Dakhan—Positions of the Pindári *darras*—Hostile outbreak at Poonah—At Nágpur—General Doveton arrives—Battle and siege of NÁGPUR—General Hardyman attacks the enemy at Jubulpore—Holkar's army defeated at MAHIDPORE—Pursuit of the Pindáras—Advance of the 3rd, 5th, and Left Divisions—Colonel Philpot moves to cut them off from Gwalior—Right Division moves down—Left Division surprises them at Bechi Tál—General Brown attacks Rámpura—Also JÁWAD—Lieutenant Mathison—Dissolution of Pindári confederacies—Breaking up of the grand army—Left Division reinforced—Siege of DHÁMONI—Of MANDALAH—General Watson left in command—Satanwári—Siege of GARHÁKOTA—5th Division moves on Nágpur—Movements of 2nd and 4th Divisions—Colonel Adams leaves Nágpur to follow Báji Ráo—Remarkable march of Captain Rodber's troop—Adams defeats Báji Ráo near Siuni—Good service rendered by Bengal and Madras Horse Artillery—Siege of CHÁNDA—Operations of the Reserve—Amir Khán's guns taken by Brigadier A. Knox—Brigadier Arnold reduces forts in Háriána and Bikanir—Táràgarh and other forts in Rájputána taken—Siege of ASIRGARH—Force employed—Its defences—Town occupied—Batteries opened—Explosion of an expense magazine—Engineer's report on the attack—Lower fort evacuated—Progress of the attacks east and west—Capitulation—Remarks on the war—Severe work performed by the troops—Deficiency of artillery officers and of siege ordnance.

THE PINDÁRI AND MÁHRÁTÁ WAR.

UNTIL the authority of the Queen of England was re-established in India by force of arms in 1857-58, no military combinations so extensive had ever been undertaken in the peninsula of Hindustán, as those by which the Marquis of Hastings crushed out the lawless power of the Pindári robbers, and reduced four great Máhrátá states into complete subjection to the paramount power. The Pindári community was not powerful in a numerical point of view, but their trade was becoming more extensive and more popular every year, and the scarcely disavowed support of Sindiah and Holkar afforded them protection. They held some lands on the river Nerbudda, partly by gift from Sindiah, partly taken possession of by force.

It was therefore necessary for the Governor-General to place in the field a force sufficient, not only to hunt down the Pindáras wherever they might fly for refuge, but also to coerce the Máhrátá states, in the very probable event of their breaking out into open war, or, if possible, to prevent such an occurrence.

The force that was assembled was divided into two separate bodies. The one on the Bengal side, termed "the Grand Army," under command of the Marquis of Hastings, consisted of four divisions and two small columns, and numbered 43,687 men; the other, termed "the Army of the Dakhan," under Sir Thomas Hislop, commander-in-chief in Madras, consisted of seven divisions, and numbered 70,487 men—both inclusive of some small native contingents.

The Bengal army was composed of—

The 1st, or Centre Division, under Major-General T. Brown.

The 2nd, or Right Division, under Major-General Rufane S. Donkin.

The 3rd, or Left Division, under Major-General Dyson Marshall.

The Reserve Division, under Major-General Sir David Ochterlony, Bart., G.C.B.

A column under Brigadier-General Hardyman, 17th Regiment.

A column under Brigadier-General W. Toone, Bengal Cavalry.

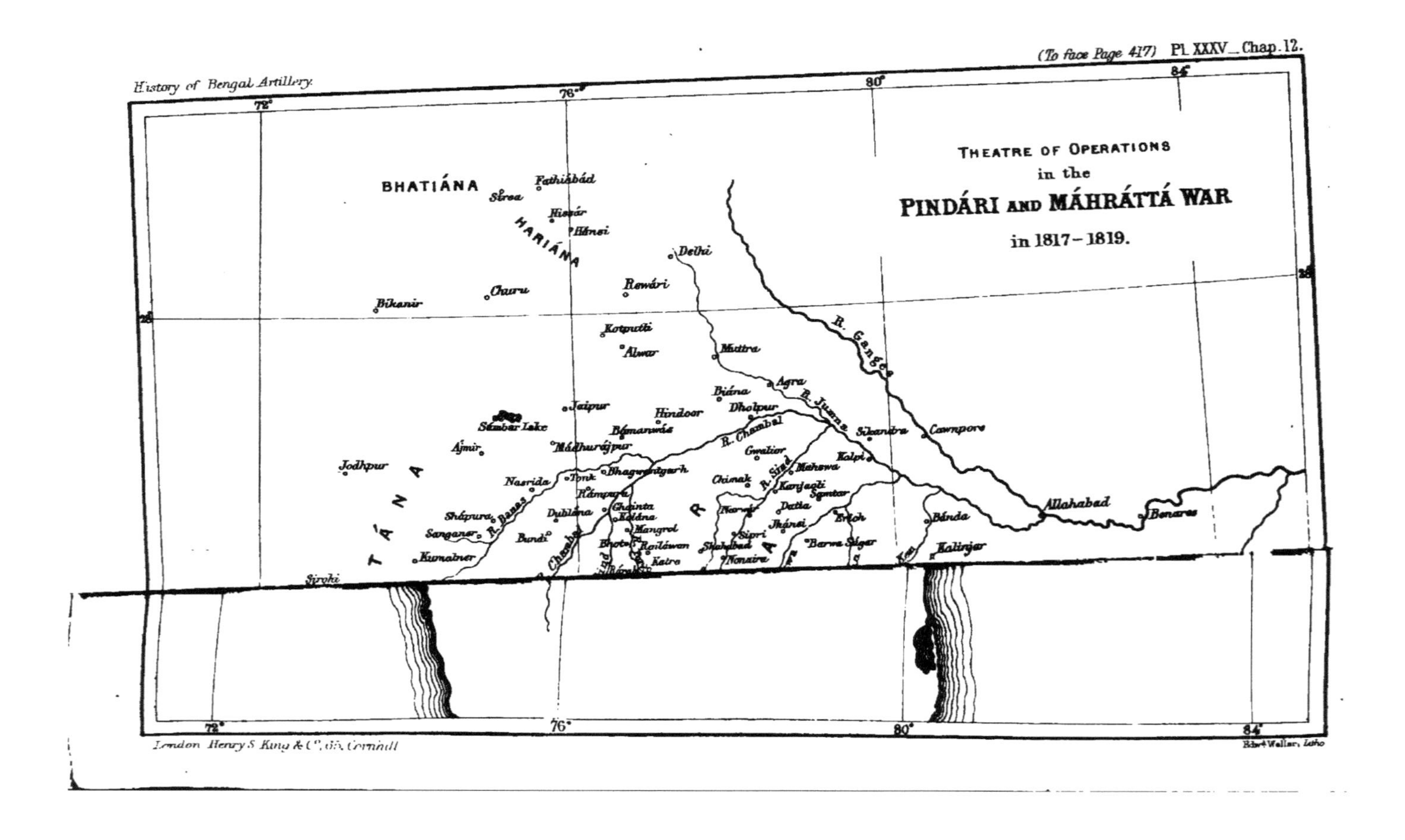

London Henry S. King & C°. 65, Cornhill
Edwd Weller, Litho

The army of the Dakhan was composed of— 1817

The 1st Division, under Lieut.-General Sir Thomas Hislop, Bart.

The 2nd Division, under Brigadier John Doveton.

The 3rd Division, under Brigadier-General Sir John Malcolm, G.C.B., K.L.S.

The 4th Division, under Brigadier-General Lionel Smith, C.B.

The 5th Division, under Colonel J. W. Adams, C.B., Bengal N.I.

The Guzárát Division, under Major-General Sir William G. Kier, K.M.T.

The Reserve Division, under Brigadier-General Thomas Munro.

It would be beyond the scope of this work to detail the movements of the whole of these different bodies; they are to be found in the narrative of this war, written by Colonel Blacker, late quarter-master general of the Madras army, one of the few books treating of Indian campaigns which may rank as a military history. It will be sufficient, therefore, to confine the present relation to those divisions to which portions of the regiment of Bengal Artillery were attached. These were the divisions of the Bengal army, and the 5th Division of the southern army under Colonel Adams, which was composed of Bengal troops. Unfortunately, besides the official documents preserved by Colonel Blacker, few records remain of this war,* and very few relative to the Bengal Artillery.

At the opening of the campaign the divisions were stationed as follows:—

GRAND ARMY.

The Centre Division, formed at Sikandra, in the Cawnpore district, moved forward to Mahewa, on the river Sind, on the 7th of November.

The Right Division, having assembled at Agra, reached Dholpur on the 8th of November.

* That is, as to details. Prinsep gives a very clear account of the war, which Captain Grant Duff makes use of in his work.

1817 The Left Division assembled at Kálinjar, and, moving by Hattah and Garhákota, reached Rehli on the 12th of November.

The Reserve assembled at Delhi not so early as the other divisions; it moved to Rewári on the 27th of November.

Brigadier-General Hardyman's column occupied positions between the Left Division and General Toone's, of which the centre was about Rewah.

Brigadier-General Toone's column, communicating on its right with General Hardyman's, took up a position at Untári, south of the river Son, on the 6th of November, and completed the northern line of operations on the left.

Army of the Dakhan.

The 1st and 3rd Divisions.—The different corps moved from the vicinities of Sikandarábád, Pandarpur, Jálnah, and Sarur, upon Hardah on the river Nerbudda, where Sir Thomas Hislop joined them on the 10th of November.

The 4th Division took up its position, extending from Ankota eighteen miles eastward. It was composed principally of regiments belonging to the Bombay establishment.

The 5th Division was formed of Bengal corps which were already stationed in the valley of the Nerbudda. It was collected at Hoshangábád by the 6th of November.

The Reserve was formed at Chinnur on the 16th of November. It had previously been employed south of the Nerbudda.

The Guzárát Division was assembled at Baroda on the 3rd of December.

On the south, a chain of posts extended from the most western point of the British frontier on the Tumbadra to its confluence with the Kistna, and thence along the latter river as far as Chintapilli, whence they were continued along the eastern gháts as far as the Chilká lake. This line, 850 miles in length, materially contributed to keep the limits of the theatre of war. The Pindári detachments, having only plunder for an object, were deterred by the smallest body of our infantry from approaching a British post.

The detail of corps composing the divisions, whose movements will now be briefly noticed, was as follows:—

CENTRE DIVISION, GRAND ARMY. 1817

Major-General T. Brown, commanding.

1st Brigade of Cavalry (Lieut.-Colonel Phillpot, 24th Light Dragoons, commanding).—24th Light Dragoons; 3rd and 7th Bengal Cavalry.

1st Brigade of Infantry (Brigadier-General d'Auvergne, commanding).—87th Regiment; 2—25th and 1—19th Bengal N.I.

3rd Brigade of Infantry (Colonel L. Burrell, 13th N.I., commanding).—2—11th, 1—24th, and 2—13th Bengal N.I.

2nd Brigade of Infantry (Colonel George Dick, 9th N.I., commanding).—2—1st, flank battalion; and 1—8th Bengal N.I.

Artillery.

Lieut.-Colonel George Mason, C.B., commanding.
Captain W. Battine, brigade-major.

*Horse Artillery.**

Major G. Pennington, commanding.
Lieutenant J. Sconce, 3rd Troop, acting-adjutant.
„ T. Lumsden, 2nd Troop, quarter-master.

1st Troop	Captain J. P. Boileau.	A—C R.H.A.
2nd Troop	Captain H. Stark.	A—F R.H.A.
3rd Troop	Captain J. H. Brooke.	B—C R.H.A.
7th, or Rocket Troop	Captain W. S. Whish.	B—F R.H.A.

Foot Artillery.

Lieut.-Colonel G. Mason, C.B., commanding.
Lieutenant H. J. Wood, quarter-master.
„ T. Timbrell, adjutant European Field Artillery.

4th Company, 2nd Battalion ...	Capt.-Lieutenant E. Pryce.	Reduced in 1826. 1—22 R.A.
6th „ 3rd „ ...	Lieutenant R. C. Dickson.	2—25 R.A.
7th „ 3rd „ ...	Lieutenant T. Timbrell.	(reduced in 1871.)

* Colonel Blacker gives the detail of ordnance for this branch as—guns, 14; howitzers, 8. The three troops had each, as then laid down, two 12-pounder and two 6-pounder guns, and two howitzers. If the gallopers of the 3rd and 7th Regiments Native Cavalry (which had not when the army was brigaded at Sikandra, been struck off the strength of their regiments) be included, the detail would be ten 6-pounder guns, six 12-pounder guns, and six howitzers, which would give the same total. But equipment is guided as much by necessity as regulation, which is born and bred in a comfortable official home; while the other leads an uncertain life out of doors. Colonel Blacker's information on this head is very meagre, and there are discrepancies in his account of the ordnance which make it very unreliable.

1817

1st and 10th Companies Golandáz, Capt.-Lieutenant T. Chadwick (4th Company, 3rd Battalion).

11th and 12th Companies Golandáz, Lieutenant J. C. Carne (4th Company, 2nd Battalion).

RIGHT DIVISION.

Major-General Rufane S. Donkin, commanding.

2nd Brigade of Cavalry (Lieut.-Colonel Westenra, 8th Light Dragoons, commanding).—8th (Royal Irish) Light Dragoons; 1st Bengal Cavalry and Colonel Gardner's Irregulars.

4th Brigade of Infantry (Lieut.-Colonel Vanrenen, 12th N.I., commanding).—14th Regiment; 2—12th, 1—17th, and 1—25th Bengal N.I.

Artillery.

Major * A. Macleod, C.B., commanding.
Lieut. T. Croxton, adjutant.
Lieut.-Fireworker C. G. Dixon, quarter-master.

Horse Artillery.

4th Troop, 2nd Brig., Ben. H.A.

4th (Native) Troop, Capt.-Lieut. G. E. Gowan.

Foot Artillery.

Reduced in 1825.

2nd Company, 2nd Battalion, Capt. A. Fraser.

LEFT DIVISION.

Major-General Dyson Marshall, commanding.

3rd Brigade of Cavalry (Colonel Newbery, 24th Light Dragoons, commanding).—4th Bengal Cavalry; 2nd and 4th Risálas,† 3rd Rohilla Horse.

5th Brigade of Infantry (Brigadier-General James Watson, 14th Regiment, commanding).—1—1st, 1—26th, and 1—7th Bengal Native Infantry.

6th Brigade of Infantry (Lieut.-Colonel Price, 28th N.I., commanding).—1—14th and 2—28th Bengal N.I.

---

* Promoted to the rank of lieutenant-colonel regimentally, 15th February, 1818.

† Squadrons.

Artillery. 1817

Captain A. Lindsay* (afterwards Captain R. Hetzler), commanding.

Horse Artillery.

5th (Native) Troop, Captain J. A. Biggs. 4th Troop, 1st Brig., Ben. H.A.

Foot Artillery.

6th Company, 2nd Battalion, Captain A. Lindsay. 4–22 R.A.

Detachment 7th Company, 2nd Battalion, Lieut. T. D'Oyly (subsequently joined). C–19 R.A.

9th Company Golandáz.

RESERVE DIVISION.

Major-General Sir David Ochterlony, Bart., G.C.B., commanding.

4th Brigade of Cavalry (Lieut.-Colonel Alex. Knox, commanding).—2nd Bengal Cavalry; and two corps of Colonel Skinner's Horse.

7th Brigade of Infantry (Colonel Huskisson, 67th Regiment, commanding).—67th Regiment; 2—5th and 1—6th Bengal N.I.

8th Brigade of Infantry (Brigadier-General John Arnold, commanding).—2—7th, 2—19th, and 1—28th Bengal N.I.; and a detachment of the Sirmur (Gurkha) Battalion.

Artillery.

Major E. Butler, commanding.

Capt.-Lieut. Charles Graham, adjutant and quartermaster.

4th Company, 1st Battalion	... ...	Lieut. C. H. Bell.	2–23 R.A.
5th Company, 1st Battalion	... ...	Captain-Lieut. J. Pereira	A–16 R.A.

BRIGADIER-GENERAL HARDYMAN'S COLUMN.

8th Bengal Cavalry. 17th Regiment. 2—8th Bengal N.I. 2 horse artillery guns. †	4 foot artillery guns, with detachment of 7th Company, 2nd Battalion, under Lieutenant T. D'Oyly.	C–19 R.A.

* It does not appear from the muster-rolls that there was a field officer, adjutant, or quarter-master with this division, which is strange. It is possible that Captain Hetzler may have been present and commanding from the first; and muster-rolls are often incomplete.

† They were the gallopers of the 8th Native Cavalry, afterward withdrawn to form part of the 5th Troop (Native) of Horse Artillery under Captain Biggs.

1817

BRIGADIER-GENERAL TOONE'S COLUMN.

Some native auxiliary cavalry.	2—4th Bengal N.I.
24th Regiment.	4 battalion guns.

FIFTH DIVISION OF THE ARMY OF THE DAKHAN.

Lieut.-Colonel J. W. Adams, C.B., Bengal N.I., commanding.

1st Infantry Brigade (Lieut.-Colonel Macmorine, commanding). 1—10th, 1—22nd, and 2—23rd Bengal N.I.

Reserve Brigade (Lieut.-Colonel Gahan, commanding).—5th and 6th Bengal Cavalry, 1st Rohilla Horse, and Light Infantry Battalion.

Artillery.

Captain J. McDowell, commanding.
Lieut. G. R. Crawfurd, adjutant.
Lieut. W. G. Walcott, commissary of ordnance.

Horse Artillery.

4th Troop, 3rd Brig., Ben. H.A.

6th (Native) Troop, Captain John Rodber * commanding.

Foot Artillery.

4-23 R.A.

5th Company, 2nd Battalion, Captain J. McDowell commanding.

The general plan of the campaign was to close in upon the Pindári encampments and hunt them down. At the close of the rains of 1817 they were in three bodies, under their three noted leaders—Chitu, whose *darra* † was between Ashta, Ichâwar, and the river Káli Sind; Karim, whose *darra* lay about Bairsiah; and Wásil Muhammad, who encamped at Gárispur, west of Ságar. From the latter a *lahbar* ‡ was sent out in October to plunder the province of Bandelkhand. It passed to the westward of General Marshall's division on its way south, and got as far as Máo Ránipur, where it was beaten off; and the Marquis of Hastings having detached a small

* Appointed to command, but did not join till January.

† A Pindári term for the collected force under the orders of a single chief.

‡ The name for a plundering expedition, pronounced nearly like *lubur*.

force under Major Cummings, 7th N.C., to cover Bánda, it retired southward without doing more mischief.

1817

The apprehensions of the Governor-General, that the Máhrátá powers would uphold the Pindári cause, were well founded. Sindiah, indeed, finding the Marquis of Hastings with the Centre Division at Mahewa, and Major General Donkin with the Right Division moving down from Agra upon Dholpur, signed on the 5th of November a treaty by which, among other provisions, the fortresses of Asirgarh and Hindia were to be given up, and a force under British management and control was to be maintained at his expense. This was the origin of Sindiah's, or the Gwalior contingent. But Máhrátá faith cannot be secured by treaties; and the Centre Division had to keep guard over this member of the family as long as a foe remained in the field capable of uniting with him.

On the same day that the treaty was signed with Sindiah, the Peshwa, Báji Ráo, whom no one ever trusted without living to repent it, attacked the Hon. Mountstuart Elphinstone, Resident at Poonah, whose small force held its ground, and repulsed in the battle of Kirki very unequal numbers. Reinforced by the 4th Division from Ahmadnagar, which, under Brigadier-General Smith, at once moved upon Poonah, the Peshwa was on the 1st of November attacked and driven from his capital, thenceforward for months to become a fugitive before the different columns detached in pursuit of him. The repulse of his force by a single battalion of native infantry under Captain Francis Staunton, with some horse and two guns of G Company, 1st Battalion Madras Artillery, commanded by Lieutenant William Chisholm, at Korygám, on the 1st of January, was the most brilliant occurrence of this war. His defeat

Reduced in 1825

1817 at Ashti, on the 20th of February, by General Smith's cavalry, and at Siuni by Colonel Adam's force, on the 17th of April, mainly contributed to his overthrow. Twice driven northward, twice southward, and once to the east, he maintained this fugitive warfare till his hopes failed; and he surrendered on the 3rd of June, to receive from Sir John Malcolm, as a reward for his atrocious perfidy and a life which was a disgrace even to the morals of his race, a pension of £100,000 per annum—a lenity which permitted him, at Bithur, to give himself up the better to the filthy debauchments of an Oriental prince, and bore its bitter fruit in the most painful events of 1857.

November It was not long before the example set by Báji Ráo was followed at Nágpur. Ápa Sáhib was so evidently preparing for hostilities, that the Resident, Mr. Jenkins, prepared to defend his post, and requested Colonel Adams to detach a force south of the Nerbudda, to remain there at his disposal. Accordingly, Lieutenant-Colonel R. Gahan marched on the 12th of November, from Hoshangábád, with the following force:—Two horse artillery guns of the 6th (Native) Troop, under Lieutenant G. Blake;* three troops 6th Native Cavalry; 1st Battalion 22nd Bengal N.I., with its two battalion guns—and halted at Sindkair. Another despatch from Mr. Jenkins was received on the 20th, as matters had become much more serious, and he marched for Nágpur at once, reaching Baitul on the 24th, and on the 27th Pandurna, a total distance of 126 miles.

* These were, Blacker says, the gallopers of the 6th Native Cavalry, which, with those of the 5th Cavalry, formed the 3rd Troop of Native Horse Artillery, ordered to be raised by G. O. 21st July, 1817. Lieut.-Fireworker G. Twemlow had charge of the two latter at Hoshangábád. The horses and men were ordered to be struck off the strength of their respective regiments from the 1st of December.

Pressing letters having arrived from Nágpur the night before, he here left his baggage with four companies of N.I., went on with the rest the same evening to Umri, 26 miles, and reached Nágpur at 4 a.m. on the 29th, altogether 176 miles in nine days.

1817 November

The troops at Nágpur had only consisted of two regiments of Madras N.I., three troops of the Bengal N.C., and four guns, with a detail of E Company, 2nd Battalion Madras Artillery, under Lieutenant J. Maxwell.* This small force, rendered smaller by much sickness, had been attacked on the night of the 26th of November and on the following day. The gallantry displayed by the force, the charge of the three troops of the 6th Bengal Cavalry under Captain Charles Fitzgerald, which entirely turned the day, and saved the rest from serious disaster, are well known. It was called the battle of Sitábaldi.†

Reduced in 1825

The repulse seemed to have its effect upon the Rájá, whose cowardly nature could only follow upon the track of an undoubted success, and the arrival, on the morning of the 29th, of Lieut.-Colonel Gahan's detachment enabled the force to rest without fear of another attack. A second reinforcement under Major Pitman, consisting of two battalions and some cavalry of the Nizam's army, arrived from the vicinity of Amráwati on the 5th of December; and on the 14th General Doveton, with the 2nd Division, also joined. He had received at Jaf-

December

* Subsequently promoted to the rank of captain from the 15th of April, 1817.

† The loss of the British in this action (367, including 15 European commissioned officers, killed and wounded) exceeded one-fourth of the number of fighting men under arms. Lieutenant Maxwell, commanding the artillery, was wounded, but slightly. In the general order published on this occasion by General Sir T. Hislop, dated Camp Gunnye, 14th December, 1817, Lieutenant Maxwell is recorded as one of those "who, with the most honourable zeal, have been so fortunate as to benefit by the favourable occasions which presented themselves during this arduous struggle."

1817 December

farábád, on the 20th of November, a despatch from Mr. Jenkins, informing him of the condition of affairs at Nágpur, and had marched at once.

The following officers of the Madras Artillery were with General Doveton's force:—

Lieut.-Colonel ...	J. Crosdill.
Captain (Bt.-Major)	G. J. Goreham.
,,	A. Weldon, commissary of stores.
,,	G. W. Poignand, brigade-major.
Lieutenant	P. Montgomerie.
Lieut.-Fireworker	A. F. Coull.
,,	J. M. Ley.
,,	E. King.

Ápa Sáhib received his ultimatum on the morning of the 15th. It was to surrender his guns, disband his army, to cede the valley of the Nerbudda and certain districts, to place the management of his revenue in the hands of the Resident, and a contingent of cavalry under British officers, and to give himself up as a hostage for the due observance of these terms. But it was not to be expected that the Rájá, though he might consent, would fulfil them at the hour named, viz. four o'clock next morning. The usual evasive correspondence ensued; so General Doveton in the evening beat to arms, and took up a position close to and on the right of the Residency, a mile and a half from the enemy's outposts. On his right, Lieut.-Colonel Gahan commanded the cavalry brigade—the 6th Bengal and 6th Madras Cavalry—with the Madras Horse Artillery under Lieutenant Poggenpohl, supported by the Bengal Gallopers under Lieutenant Blake. Next to the cavalry, Lieut.-General Macleod commanded a wing of the Royal Scots (1st Royals); three battalions and two flank companies native infantry, next. Lieut.-Colonel Mackellar's brigade consisted of a company of the Royals, a battalion of native

(To face Page 427) Pl. XXXVI _ Chap. 12.

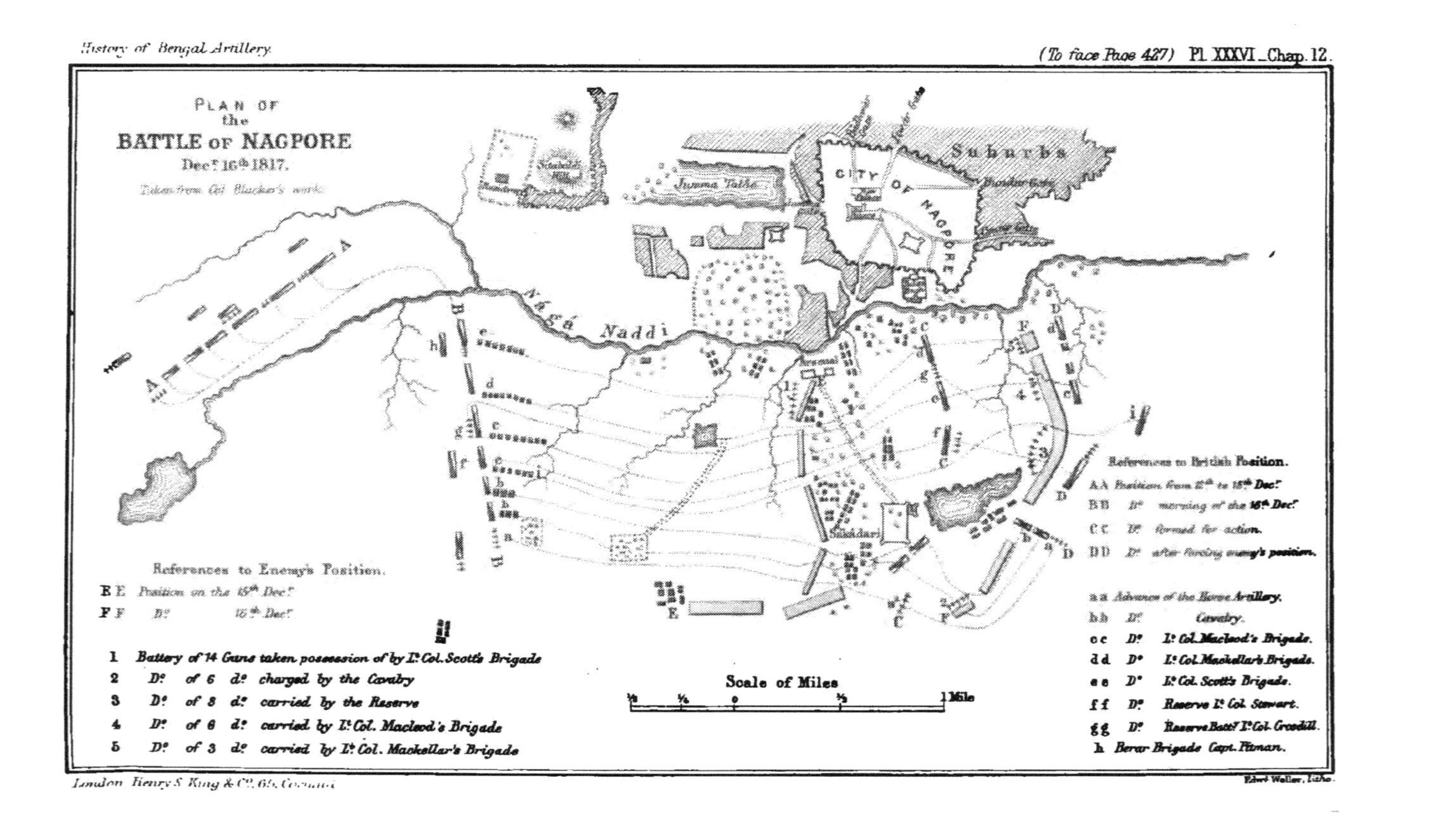

London Henry S. King & Co. 65, Cornhill

Edwd Weller, Litho.

infantry, with two horse artillery guns under Lieutenant Hunter, and two field-guns, also of Madras. Lieut.-Colonel Hopeton Scott's brigade, on the left, consisted of a company of the Royals, a battalion and a company of N.I., the sappers and miners, and four guns with a detachment of Madras Artillery. Lieut.-Colonel Stewart commanded a battalion of native infantry as a reserve; and Lieut.-Colonel Crosdill another reserve of infantry, and a battery of guns commanded by Captain Weldon, in rear of the right infantry brigade.

1817 December

In the morning, nothing having been done to carry out the agreement, General Doveton moved forward and took up the position *B B*, when notice was brought that Ápa Sáhib had surrendered to the Resident. After more evasion and delay, it was reported that the guns were to be handed over at 12 o'clock.

See Plate XXXVI.

The position of the enemy was secured in front by extensive suburbs, buildings, plantations, and avenues of trees; the flanks of both armies towards the city were secured by the Nágá Naddi, a stream which skirted its southern side, and from which three ravines, running parallel, intersected the ground between. On the other flank and beyond, the country was open and unobstructed.

The general, on questioning the messengers, suspected that the guns would not be quietly given up, and therefore advanced his whole line. On entering the plantation, a battery of 14 guns (1) was taken without resistance,* together with about 23 more in an arsenal close by; but a sharp musketry fire was opened upon the right from an inclosed garden, and the line deployed when within range of the other batteries, which now began,

* A lascar was in the act of applying a slow match to one of the guns as they came up, but he was seized.

1817 December

and kept up an incessant fire in front and on the right. The cavalry and guns under Lieut.-Colonel Gahan moved round outside the Sakádari garden and suburbs, leaving a reserve of 100 men from each regiment and the horse artillery in rear. They charged and captured two batteries successively, which were supported by a strong body of horse and foot. Their isolated position encouraged the enemy to renew the attack; but the horse artillery, with an effective fire, checked the attempt, and enabled the cavalry to continue the pursuit.

Meanwhile the infantry brigade, advancing, charged and took the batteries opposed to them (3, 4, and 5). Two of these on the left fell to Lieut.-Colonels Macleod and Mackellar; the centre one, which from its position had been the most destructive, to the reserve under Lieut.-Colonel Stewart. The fire of the artillery, under Lieut.-Colonel Crossdill, "on both occasions materially contributed to the successful result."[1] At half-past one o'clock the whole of the batteries and the enemy's camp left standing were secured. Our loss in killed and wounded was 141; most of the latter being by cannon shot, many deaths resulted. In the artillery, the casualties were only one gunner of the horse artillery in Lieut.-Colonel Mackellar's brigade wounded, and 8 horses killed or missing.*

The enemy, though they had abandoned their ground and their guns, had not yet given in. Man Bhat Ráo, with the Arab infantry, of whom there was a large number, held the city and refused to evacuate. The centre part of the city was enclosed by a wall of the usual kind, only not more than eight feet high, flanked

* The following artillery officers were honourably mentioned in General Doveton's despatch, dated 19th December:—Lieut.-Colonel Crosdill; Major Weldon; Lieutenants Poggenpohl and Hunter, Madras Establishment.

[1] Blacker, p. 127.

at intervals by round towers, and surrounded by extensive suburbs. Within, well-built houses and narrow lanes offered the best opportunity for street fighting, a species of defence in which the Arabs excelled. It was necessary therefore to proceed cautiously. As the battering train was still at Elichpur, it was necessary to utilize such of the captured guns as were considered serviceable. Including these, the ordnance employed amounted to—

1817 December

5½-inch howitzers (two heavy)	... ...	3
6-pounder guns	... ...	7
7 " "	... ...	4
12 " "	... ...	1
15 " "	... ...	1

Some sand-bags constituted the whole of the engineer's stores, and the line regiments were indented upon for entrenching tools.

The Jumma Darwáza was the first point selected for attack, as it was intended from there to batter the old palace, the most commanding point within. A lodgment was made on the Jumma Taláo, an extensive tank, or rather lake, on the 19th of December, where a howitzer battery was erected; and an advance was made to the eastern side next day, but slowly, owing to the backwardness of the native pioneers. A battery for four guns was, however, completed on the eastern side, and during the night another for five of the enemy's guns was commenced on the same side, to its left. Nothing was done on the 21st, in consequence of some negotiations having been opened; but during the two following days the fire appeared to have rendered the breach at the gate practicable, and it was stormed next day. The assaulting party, however, though they attained the breach, could not go further, owing to the destructive fire kept up from the buildings within, nor could they find cover elsewhere; so they were recalled.

1817 December

The loss on our side during the siege, from the 19th to the 24th, amounted to 307, of which the artillery bore the following share:—Killed: 4 gunners. Wounded: Major Goreham, slightly; Lieutenant Coull, severely; 2 corporals, 13 gunners, 1 havildár, and 13 golandáz.

Notwithstanding the bad success of this attempt, the enemy, who held the town, did not consider their position worth fighting for, and they again, the day following, renewed their offers to evacuate upon terms which were agreed to. With their arrears paid up and a gratuity, they were allowed to leave the Nágpur dominions, to offer their swords to some other power better able to carry on the war against Feringhi control. The concession was wise; it completed the disarmament of one member of the Máhrátá confederacy, and set free a great portion of the force under General Doveton, for service elsewhere.

When the first rumour of the rupture with the Bhonslah reached the Governor-General, he ordered Brigadier-General Hardyman to move down upon the Nerbudda, and there wait for any orders he might receive from Mr. Jenkins, at Nágpur. All along the frontier of the Bhonslah's dominions his officers were organizing their forces to aid the general rising. Major Richards, commanding a small detachment at Jubulpore, found himself under the necessity of abandoning that place and falling back upon Lieut.-Colonel Macmorine, who, with a small part of the 5th Division, was at Garawára. The latter officer, as well as Major Macpherson, at Hoshangábád, had been obliged to call in his outposts, which were too scattered to afford one another effective support. On the 20th Macmorine, continuing the retrograde movement, had joined Macpherson at his station.

Brigadier-General Hardyman moved down upon

1817 December

Jubulpore, where a large body of the enemy had posted themselves, to the number of 2000 infantry and 1000 cavalry, with some guns. He had with him the 17th Regiment, commanded by Lieut.-Colonel C. Nicol, the 8th Bengal Cavalry, and four guns under Lieutenant T. D'Oyly. A native infantry regiment which had not time to come up was halted at Belléri. His route lay through Badanpur, Bellári, and Talwáh. On the 19th he found the enemy occupying a rocky eminence in front of the fort and town. Sending Major Lucius O'Brien * with two squadrons round to the right, to gain their rear and cut off their retreat by the river, he advanced with his guns in the centre of the infantry, and masked by a portion of the 8th. On nearing the enemy he uncovered his guns, which plied them with shrapnel for a quarter of an hour, when the reserve squadron of the 8th charged and carried the guns in front.† Then the infantry went forward and took the heights on either flank, whilst the advanced squadron went round the hill by the right and pursued the fugitives as they were dislodged from above. They fled, leaving nine guns and many stores behind. After this success the general moved forward as far as Chaprah, where, on the 26th, he met a despatch from the Resident, dispensing with his services, and he returned to Jubulpore.

There was yet the army of Holkar, which in the

* The same officer who had commanded the Java Cavalry and troop of horse artillery attached, alluded to at the concluding part of chapter ix.

† The officer who led this charge was Lieutenant Alexander Pope, who afterwards commanded the cavalry on the right at Chiliánwálá. General Hardyman, in his despatch dated the 20th December, thus mentioned him :—

"It fell to the lot of one individual to be more conspicuously distinguished than the rest, and that officer is Lieutenant Pope, of the 8th Cavalry. He charged steadily under a heavy fire from the heights, penetrated to the enemy's guns, received a spear into his body, and continued the pursuit with vigour."

1817 December

beginning of December had not taken any decided measures. The chief of that name, Malhár Ráo, was only twelve years old, and his stepmother Tulsi Bái, who to the usual characteristics of a Máhrátá added the profligacy of a Messalina, was regent. But the army was mutinous, and determined to take up the cause of Báji Ráo, from whose treasury they hoped to realize the arrears of pay due to them, while the regent was secretly making terms with the English. The army had left Rámpura, on the Chambal, and was moving slowly towards the Dakhan. Sir John Malcolm, who with the 3rd Division was following up the Pindári chief Chitu from Talain, heard that he had been received into the Máhrátá camp, and therefore fell back upon Ujain, where on the 12th he joined the 1st Division, under Sir Thomas Hislop.

A movement was immediately made towards the army of Holkar, at this time encamped at Mahidpore, on the river Siprah. In the early morning light of the 20th of December, a party of soldiers conveyed down to the banks of this stream a covered litter, from which a woman was taken out and put to death. It was Tulsi Bái, whose secret negotiations were more than suspected. But the Máhrátá and Pathán soldiers of Malhár Ráo Holkar had other work for their swords next day. Sir Thomas Hislop, passing the Siprah by a single ford in front of their army, and well within cannon range, attacked and completely routed them in the most decisive action of the war. The light horse artillery guns were no match for the heavy pieces they had to fight, and were disabled. They had five rank and file, 3 officers' chargers, and 35 horses killed; eight horses missing; and Lieutenants Gamage and T. G. Noble, Troop Quarter-master Griffin (the latter severely), 1 staff sergeant, 1 sergeant, and 10 rank and file wounded. The enemy were dispersed, their camp taken, and 63 out of 70 guns captured. Lieutenant F. S. Sotheby, of the Bengal Artillery, was present in this action with the guns of the Russell Brigade, to which he was then attached.* The loss of the British was very severe. But Holkar's power was broken, and his army disappeared as a body from the theatre of war.

1817 December

The Centre Division before this had met a different enemy. Just after the signature by Sindiah of the treaty presented to him by the Governor-General, it was attacked by cholera in a very severe epidemic form The camp was moved about as the disease permitted and finally remained from November 20th to December 5th at Erich, on the river Betwa, where it finally disappeared.

November

The artillery fortunately suffered but little. By the muster-rolls dated December 1st, it appears that no deaths were reported in the 1st, 2nd, and 3rd Troops of Horse Artillery; in the 4th Company, 2nd Battalion, a sergeant and four matrosses; in the 7th Company, 3rd Battalion, a gunner and three matrosses died during the month of November.

Meanwhile the Left Division, with the 3rd and 5th, had commenced work. It has been said that the three Pindári *darras* were encamped between Ságar and the river Parbatti. On the 11th of November General Marshall had advanced from Kálinjar to Raili. The 1st, 3rd, and 5th Divisions were on the banks of the Nerbudda, prepared to move northwards; the Guzárát

* The following Artillery are honourably mentioned or thanked in Sir T. Hislop's despatch of the 22nd:—Major Noble, C.B., commanding the arm; Captain Rudyerd, commanding the Horse Artillery and Rocket Troop; Lieutenant Bennet, commanding a Field Battery of the Madras; and Lieutenant Sotheby, Bengal Artillery, commanding the guns of the Russell Brigade.

1817 December

two *darras*. It was a success, but not a complete one. The artillery were not engaged here, as they were being dragged up the pass when the cavalry went on ahead.

Next day, the 16th, General Donkin, at Kolána, finding the enemy driven up towards him, pushed forward with his light brigade by night, and caught up their baggage early next morning. The family of Karim Khán were among the captures. Colonel Adams's cavalry, under Major Clarke, fell in with some more of them, and cut them up. Shortly after, as the colonel was moving again southward, he detached Captain Roberts with the 1st Rohilla Horse against a party of 400, who were nearly all destroyed.

1817-1818 Dec.—Jan.

Major-General Marshall, with the Left Division, after surprising the enemy at Bechi Tál, had advanced to Bárah, beyond the river Parbatti (December 16th to 18th), from whence he was ordered to return to Sironj, which he reached on the 4th of January.

Colonel Adams countermarched from Katra, and moved to the south-west by Bároda, Eklera, Gugarni, Kotra, Susner, to Gangrár (December 20th to January 6th), where he halted eleven days.

The Guzárát Division, under Sir W. G. Keir, was also on the move from the south, to co-operate against the Pindáras. His route from Dháwad was by Patláwad, Ratlám, and Mundisor (December 19th to January 3rd).

The Right Division returned from the banks of the Parbatti to Ghainta Ghát, on the Chambal (December 19th to 21st), whence, after a week's halt, he moved westward by the Bundi pass to Sánganer, behind the Banás river, to guard the northern outlets of the country within which the Pindári detachments were now enclosed.

A force equipped for light movements was placed under the orders of Major-General T. Brown, for the purpose of aiding in the pursuit. It consisted of the following corps:— 1817–1818 Dec. Jan.

From the Centre Division.

3rd Bengal Cavalry.
Dromedary Corps.
Two 12-pounder guns.
2nd Troop Horse Artillery, under Lieut. Matheson.

From the Left Division.

4th Bengal Cavalry.
Three risálas of Cunningham's Horse.
1st Battalion 18th N.I.
5th Troop (Native) Horse Artillery, under Captain Biggs.

Major-General Brown left the Governor-General's camp at Sonári on the 18th of December by Chimak, Narwár, Sipri, Deori, to Nonaira (December 18th to 25th), where* the detachment from the Left Division, then only a few miles off (at Bechi Tál), joined him. Thence he marched by Naharghar, Baroda, Ratlái, Pachpahár, to Pipliah (December 26th to January 9th), where he heard that Roshan Beg and Roshan Khán, the two chiefs of the Pathán soldiers of Holkar's army, were at Rámpura, twenty miles further on. He appeared before the place early the next morning with the 3rd Cavalry, the dromedary corps, and some native infantry. The principal part of the enemy had left the place, and the rest made no defence, but lost more than half their number and all their baggage. Shortly after, a number of their guns were given up by the chief of Ahmadgarh about ten miles from Rámpura.

* This is on the authority of Colonel Blacker, who, though he does not give the name of the place where the detachment joined, puts it before General Brown's arrival at Bechi Tál, where the Left Division was on the 25th, 26th, and 28th December. Both columns must have met, according to Colonel Blacker's dates,[1] at Dekoodnee (Dhakwáni on the Indian Atlas), 4½ miles west of Bechi, on the 27th.

[1] Route map, No. III.

1818 January

The rapid march of Sir W. Keir from Mandisor to the west and north-west of Neemuch (January 4th to 23rd) left the flying enemy little peace, though they always evaded a fight; and Chitu made his escape with about 5000 followers into the jungles of Bánswára, south-west of Mandisor. The other two Pindári leaders crossed the Chambal on their way east, and encamped at Garária, a village a few miles north of Gangrár, where Colonel Adams had halted. Adams's information being good, he lost no time in sending Major R. Clarke with the 5th Bengal Cavalry thither, who, arriving unseen close to their camp fires before daylight on the 13th of January, posted his men so judiciously, that the estimate of 1000 slain out of 1500 is said not to be an exaggeration.

Meanwhile, General Brown had moved to Aorah, on the Chambal, where he halted on the 17th and 18th. Jaswant Ráo Bháo was still at Jáwad, and as he could not be brought to give up the Pindári leaders in his camp, the general appeared before the place on the morning of the 25th. On the 29th a squadron of cavalry was sent down to reinforce the picquets, as it was said that the Bháo was preparing to run for it. This squadron was fired upon by the enemy encamped outside the town, and the general at once turned out, sending a detachment of cavalry under Major Ridge, 3rd B.C., to watch the other side. Lieutenant Matheson, with his two guns, and two more of the native troops were sent down to support the picquets. The fire of these guns, particularly of the 12-pounders with shrapnel, mainly contributed to drive the enemy within the walls. During this part of the action one of the gunners was mortally wounded, and as his comrades attempted to take him to the rear, he begged them not to do so, adding, "I know I must die, and I only wish to shake Lieutenant Matheson by the hand before I die." His wish was gratified, and his last words to his officer were, "God bless you!" Why are such instances of soldierly brotherhood between officers and men so rare?

General Brown at once moved up to the town, and as the Bháo declined to open his gates, Lieutenant Matheson did it instead. A storming party was formed, and under cover of the infantry, supported by Captain Biggs's 6-pounders, it advanced with one of his guns. The third discharge carried away the fastenings, and the party forced their way in. Jaswant Ráo, with 100 men, escaped by an opposite gate; but the cavalry, which had only rested two hours after returning from a 25 miles' march, were too fagged to keep up the pursuit. The loss of the enemy in the attack on the town and their detached camp in the rear was about 1000 men, while that of the British was only 36. Of this, the artillery had one gunner killed and four wounded. Captain Biggs and Lieutenant Matheson were thanked by the major-general in orders, and the latter was honourably mentioned in the Government general order. Lieut.-Colonel Pennington, commanding the horse artillery, also addressed Lieutenant Matheson personally in the following terms:—

"Accept my best thanks for the great credit you brought the horse artillery by the ability and gallantry you displayed in the attack on Jaswant Ráo Bháo and his town, and my cordial congratulations on your personal safety."

This was the last service performed by Major-General Brown's detachment. He left two of Captain Biggs's guns at Jáwad, and marched with the rest to Rámpura, whence he rejoined army head-quarters.

"His detachment," says Colonel Blacker, "came unexpectedly into the theatre of operation, acted a brilliant part, and vanished again as suddenly. Its motions and effects were like those of a

1818 January

rocket among a body of enemies, and prepared the Bháo's and Bápu Sindiah's troops subsequently to surrender the moment they were summoned."[1]

Colonel Adams's Division marched from Gangrár on the 17th, leaving some infantry and part of the artillery at Ágar. Next day, he proceeded by Sarangpur, Shujáwalpur, Gurára, Islámnagar, to Pagnesir on the river Betwa (January 18th to 27th), an average of 18 miles a day. The Pindáras who had gone towards Bhopál, finding little rest, were not disinclined to accept the liberal terms now held out to those who forsook their lawless trade, and one by one came in and submitted.

Chitu, meanwhile, had managed to avoid a direct collision with our troops, but everywhere the country was becoming too hot for him. Eluding the Guzárát Division, he crossed by a very difficult pass over the hills near Dhár into the valley of the Nerbudda, losing his baggage and many of his followers. Pursued and plundered now by Bhils and Grásiyas—tribes that could claim a far greater antiquity in the robbers' trade than any Pindára—the submission of his followers completed the prime object of the war. But Chitu declined to give in, preferring to wander about a fugitive outlaw, till in February, 1819, he yielded to a tiger, in the jungles of Satwás, the life which he would not trust in the hands of a British officer.

February

The continuance in the field of the whole force being now unnecessary, orders were given for the breaking up of the Centre Division. A column was placed under command of Brigadier-General J. Watson, 14th Regiment, to assist the Left Division in the reduction of places along the Nerbudda. Its movements will be mentioned hereafter. The horse artillery and rocket troop were ordered to Meerut.

[1] p. 210.

(To face Page 442) Pl. XXXVII _ Chap. 12.

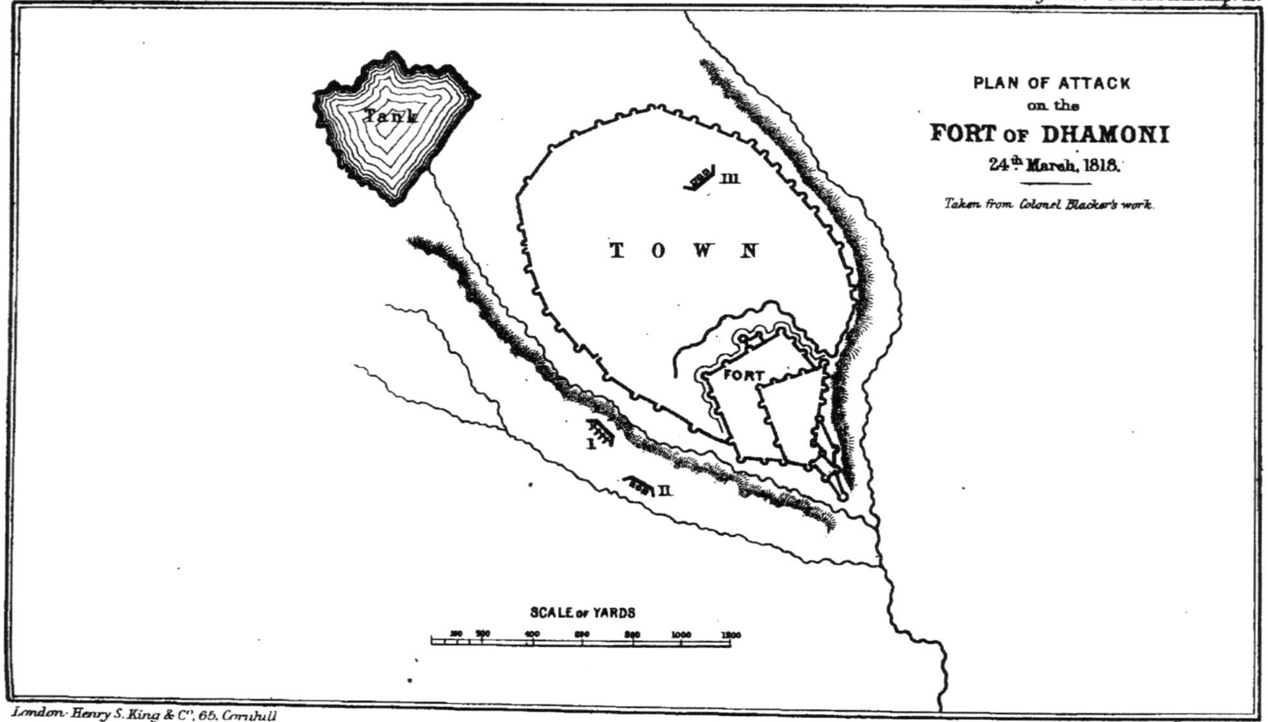
PLAN OF ATTACK
on the
FORT OF DHAMONI
24th March, 1818.
Taken from Colonel Blacker's work.
Tank
TOWN
FORT
I
II
III
SCALE OF YARDS
100
200
400
600
800
1000
1200

London. Henry S. King & Co. 65, Cornhill

Edw. Weller, Litho.

Major-General Donkin's Division was also broken up. The foot artillery were ordered to Agra, which they reached on the 27th of March, except four guns with golandáz, left at Tonk Rámpura, under Lieutenant Croxton. Captain Gowan, with his troop of horse artillery, went by Bhámpura to Jáwad, where the 3rd Cavalry had been left, and subsequently formed part of a force called the Rájputána Field Force, under Colonel J. Ludlow.

1818
March

Brigadier-General Watson's column was composed of the following troops:—2nd Battalion, 1st N.I.; 2nd Battalion, 13th N.I.; 1st Battalion, 26th N.I.; 7th Bengal Cavalry; 4th Company, 2nd Battalion Artillery; and the battering train. With this force he left Kanjáoli and proceeding by Samtar, Barwa-Ságar, Tehri, Áston, Sindwáha, and Málthon, reached General Marshall's division at Khemlása on the 5th of March. The duty which devolved upon this force was to take possession of the many forts in the Ságar district and Nerbudda valley, which had been made over by Ápa Sáhib in the treaty just concluded with him. Several of these places submitted to small detachments sent out, but others had to be reduced by force.

February

March

One of the latter was Dhámoni, a town 24 miles north of Ságar, and thither General Marshall proceeded with part of his force, arriving before it on the 19th of March. The fort, pentagonal in shape, is situated on an eminence at the eastern extremity of the town, which is on the edge of a ravine, running into the river Dassán. It had an interior line of defences, and the Kiladár's house was also fortified. Possession was taken of the town, the wall of which was in a ruinous state, without difficulty. Materials having been collected, a breaching battery was erected on the night of the 23rd, for two 24-pounders and

PLATE NO. XXXVII.

1818 March

four 18-pounder guns, as well as another on its right for a 12-pounder gun and two howitzers, while a mortar battery (No. II.) was constructed in rear. Another mortar battery (No. III.) was placed in the town. On the morning of the 24th the batteries opened, and kept up an incessant fire for six hours, when the Kiladár surrendered unconditionally.

The value of vertical fire is beyond dispute in reducing such places. If proof had been wanting, it was found at Háthras, and General Marshall had not forgotten it: otherwise, judging from the plan, the western face should have been attacked. The ridge on which Batteries No. I. and II. were placed was selected on account of its having a good command and affording good cover. But as a general rule, the front attacked should have been also enfiladed, which was not done, though the town (which was in our possession) must have afforded cover for the purpose. This, however, is merely judging from a bare plan. The artillery officers here present were Captains Hetzler and Lindsay; Captain-Lieutenant Coulthard; Lieutenants Carne and W. Bell; Lieutenant-Fireworkers Sanders, Crommelin, and Patch.

April

The force next proceeded against the town of Mandalah, on the river Nerbudda, by Kythaura, Sanodha, Sháhpur, Dumoh, and Jubulpore. Major Lucius O'Brien, while acting as commissioner to take over ceded forts, etc., had been treacherously attacked, under secret orders from Ápa Sáhib, by the Kiladár of this place on the 1st of March, an act which called for punishment. It was reached on the 18th of April, but it was not for nearly a week that the whole of the store carts had come up; the country, intersected with ravines and hills, being very difficult.

PLATE No. XXXVIII.

Mandalah is situated in the angle of a sudden bend

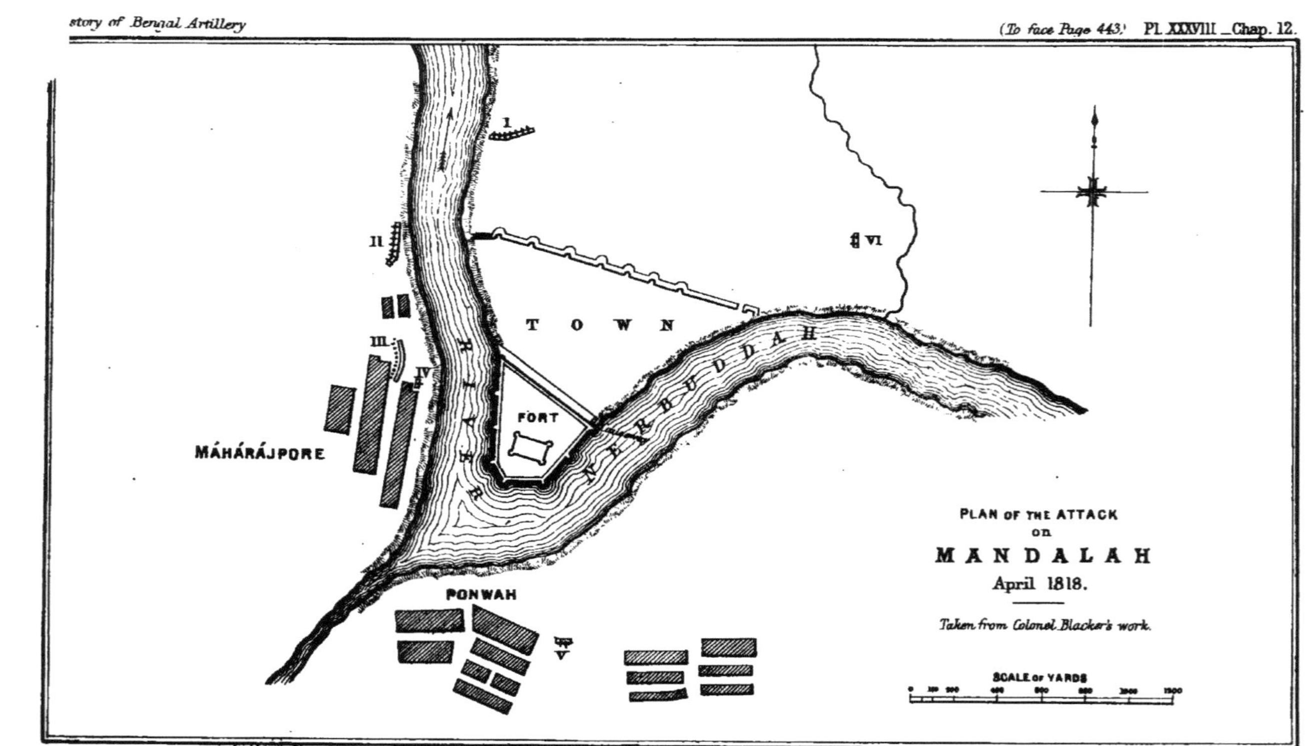

London: Henry S. King & Co., 65, Cornhill.
Edwd Weller, Litho

in the Nerbudda, on its right bank; the fort being at the apex of the triangle. It was separated from the town by a broad cutting, which converted it into an island, connected with the mainland by a causeway at its eastern end. The whole of the works were liable to be taken in reverse.

1818 April

To reduce it the following batteries were erected:—

No. I.—Two 18-pounder, one 12-pounder, and two 6-pounder guns, to breach the west extremity from the north at 500 yards.

No. II.—To enfilade the same side and assist in breaching from the opposite bank of the river, here about 350 yards wide; two 24-pounder and two 18-pounder guns; two 8-inch and two 5½-inch mortars.

No. III.—Close to the village of Máhárájpur, to the right of No. II.; three 10-inch, three 8-inch, and eight 5½-inch mortars.

No. IV.—A single 12-pounder gun, placed about 100 yards to the right of the mortars, to check the enemy's fire.

No. V.—Near the village of Ponwáh, to enfilade the causeway connecting the fort with the town; two 6-pounders.*

No. VI.—Two more 6-pounders, within 450 yards of the eastern extremity of the northern wall.

With great exertions the whole of the ordnance was conveyed down (the sepoys assisting) to the batteries during the night of the 25th, and next morning opened fire. By 2 p.m., so much effect was produced on the wall that the troops intended for the attack were crossed over to the right bank. These consisted of fifteen companies N.I., under Colonel Dewar, with a reserve of thirteen under Colonel Price. Captain Tickell, the senior engineer, and Lieutenant Pickersgill, assistant quarter-master general, examined the breach; and finding it practicable, the assaulting columns, led by Brigadier-General Watson, advanced under a fire from the enfi-

* Apparently battalion guns, as they are called in General Marshall's despatch (Papers, p. 328) "Captain Black's battery." This officer (Alex. Black) was in the native infantry; he commanded the line of posts at the village of Ponwáh.

1818 April. lading battery. The breach was carried, and the garrison was driven back towards the fort. The gates of this were closed against them, and all were killed or drowned, except some fifty taken prisoners. Battery No. V. rendered essential aid while the town was being cleared. In the evening the Kiladár surrendered the fort, which was occupied by our troops on the morning of the 27th. Twenty-six guns, the largest a 68-pounder, were found inside, with abundance of ammunition.*

Captain Hetzler, as before, directed the artillery operations of the siege, with the following officers:—Captain Lindsay; Lieutenants Carne, Dickson, D'Oyly, Kirby, Sanders, Crommelin, and Patch.

Immediately after this, Major-General Marshall was ordered to return to his command of the Cawnpore Division, and made over to Brigadier-General Watson charge of the Ságar Field Force, as it was now called. It left Mandalah on the 30th of April, and, after some counter-marching, reached Ságar on the 24th of May.

May. The fort of Sátanwári, eighteen miles west of Bairsiah, was next invested by a detachment under Major W. Lamb, with a small battering train of two 18-pounders, four mortars, and two field-pieces (probably howitzers), and a detail of the 4th Company, 2nd Battalion, with Lieutenants Carne and Saunders. The detachment left Ságar on the last day of May, and reached the fort on the last day of June. During the night of the following day batteries were erected, and opened at daybreak of the 10th, keeping up a steady fire until 5 o'clock p.m., when the engineer, Captain Tickell, reported the breach practicable, and the assault was ordered. The garrison

June.

* Casualties among the artillery were one golandáz and three ordnance drivers wounded.

(To face Page 448) Pl. XXXIX _ Chap. 12.

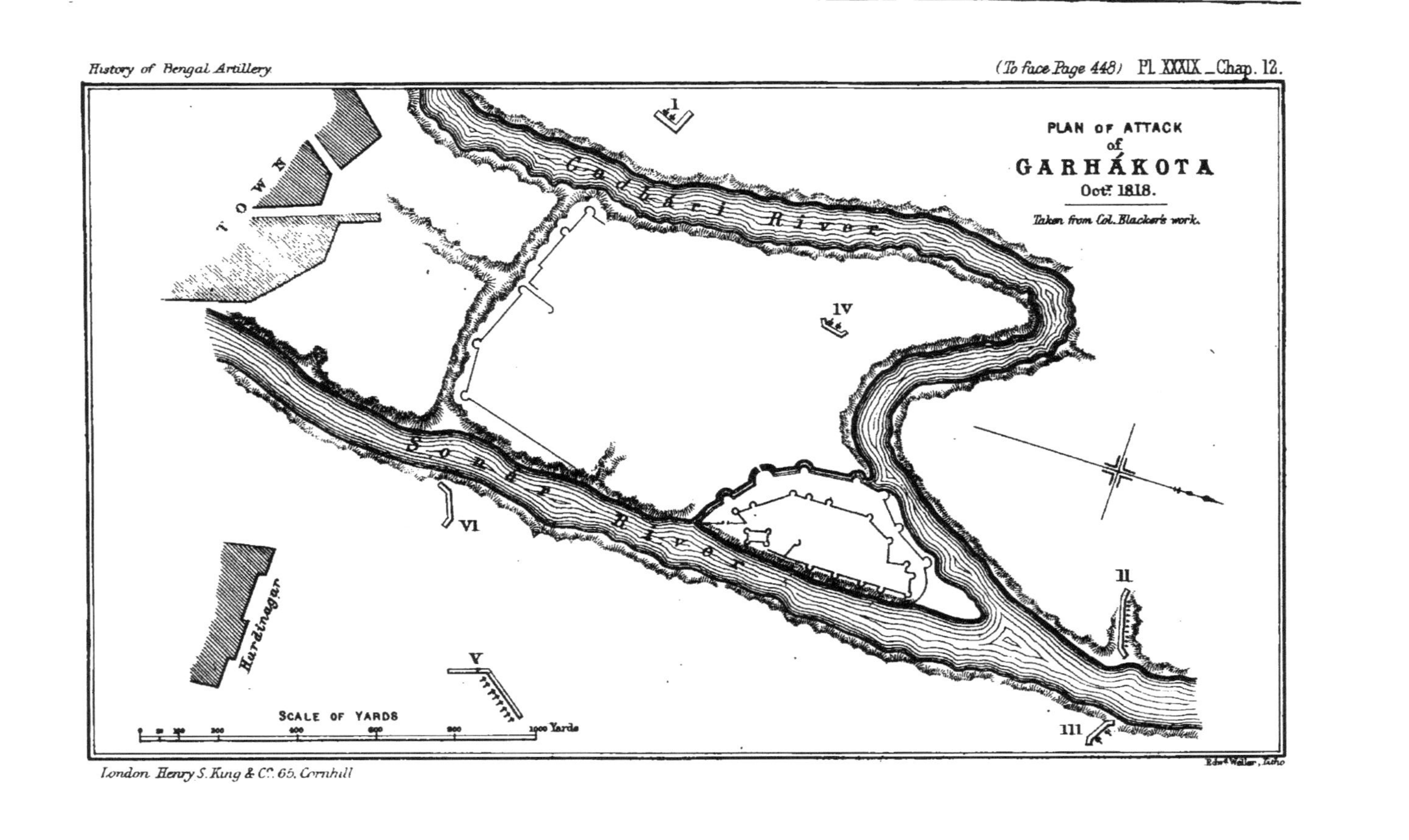

London. Henry S. King & Co. 65, Cornhill.

1818 June

however, good marksmen from the jungles,* poured in a destructive fire upon the head of the column. The sepoys in the rear, disheartened, made for the nearest cover, and no exertions of their officers could get them forward; so the attempt failed, and they had to be recalled. During the night, nevertheless, the place was evacuated; but Major Lamb having taken precautions for this, it was not without some loss that they got away.

The loss of the 4th Company in this day's work was a corporal killed; 1 sergeant, 1 gunner, 5 matrosses, 2 golandáz (attached), and 3 lascars wounded.† The detachment then returned to Ságar.

October

At the conclusion of the rains a force from Ságar was again called out. One Arjun Singh, formerly Kiladár of Garhákota, seized upon it on his own account; and Sindiah, to whom it had been made over, applied for a force to recover it. Accordingly, Brigadier-General Watson left Ságar on the 15th, and arrived before the place on the 18th of October. Captain Hetzler commanded the artillery, which consisted of—

* Baigás—called by Blacker "Baugrees"—"the wildest of all the hill tribes" inhabiting the district (till lately unexplored) about the head-waters of the Sone and Nerbudda rivers; a branch of the Gonds, one of the aboriginal families of Central India.—See the introductory chapter, "Central Provinces Gazetteer," pp. cvii., cxv.—cxvii.

† Notwithstanding that the breach had been declared practicable by an engineer who had a more than ordinary experience in siege operations, a report was spread that the breach had not been sufficiently cleared. This led to Lieutenant Carne's trial by court martial on three charges—for bad practice, for serving out an excess of liquor to the men in battery, and for being himself drunk. He was acquitted wholly of the first and third, partly of the second. When we consider the bibulous tendencies of those days, it is hardly surprising that this officer yielded to the solicitations from hard-worked men for "another tot;" but to forget the effect of the sun in June and the nature of the service he was employed on were grievous errors, for which in his arrest and trial he paid dearly.

Reduced in 1825

The 4th Company, 2nd Battalion; Captain-Lieutenant Coulthard, Lieutenant Sanders, and Lieutenant-Fireworker Crommelin.

4-22 R.A.

The 6th Company, 2nd Battalion; Lieutenant Pew (Adjutant) and Lieutenant-Fireworker Counsell.

Medical Staff-Surgeon C. Campbell and Assistant-Surgeon H. Smith.

PLATE No. XXXIX.

The town and fort are situated on the angle formed by the confluence of the Gadhári and Sonár rivers, and are above a mile apart from one another. A wall running across the tongue of land on which they are placed, separated both at about 1600 yards from the ditch of the fort. The latter had a double *enceinte*. General Watson obtained possession of the town without difficulty on the 18th, and the same night erected a battery (No. I.) to rake the wall above mentioned. On the night of the 23rd, a battery (No. II.) for fourteen mortars and four howitzers was placed on the left bank, below the junction of the two streams, about 1000 yards from the north angle of the fort; and on the opposite bank two 6-pounders (No. III.) were placed, to enfilade the interior of the outer wall along the Gadhári. On the 24th two more 6-pounders (No. IV.) were placed, to enfilade the south face. A breaching battery of two 24-pounder, four 18-pounder, and two 12-pounder guns (No. V.) opened on the 26th, at 900 yards. An accident occurred in the mortar battery on the morning of the 24th; about 100 filled shells, placed in the rear in readiness for use, and covered with tarpaulins, exploded, it is said, by the premature bursting of one just fired. This accident cost the lives of several men in battery at the time.* On the

1818 October

* The officer in command of the battery at the time appears to have been Lieutenant Thomas Sanders, at least to judge from a story in the *East India United Service Journal* (vol. ix. p. 362) entitled "Bukheira," the humorous style of which appears to point out the late General Augustus Abbott as its author. In alluding to this accident, the writer says that Captain Thomas, as he calls him, in order to avoid the fragments of the shells, had in despair thrust himself into the muzzle of a *loaded* 24-pounder.

29th, two 24-pounder guns (No. VI.) were placed to the left front of No. V., on the bank of the river, to improve the breach, already sixty feet wide at the top; and the orders for storming next morning were issued. But the garrison had enough of it, and surrendered during the day. Their fire all along had been contemptible, and it seems at first sight curious that they should not sooner have thought of capitulation if they did not intend to fight. They lost a fifth of their number during the siege.

1818 October

Here, again, the value of a heavy vertical fire is seen. There is not the same objection to it, in the case of forts, that can be urged when a populous town is to be attacked. The two enfilade batteries, Nos. III. and IV., being directed on opposite sides of the fort, would not, on large fronts, have supported each other; but this disadvantage was not so apparent in a small work. There were too few artillery officers—only two captains and six subalterns—while there were in batteries Nos. II., III., IV., and V., thirty pieces of ordnance in action at one time; besides which, the duties of ordnance stores, staff-office work, and general supervision had to be attended to. To the want of sufficient supervision undoubtedly must be attributed the accident of the 24th of October. Inclusive of non-commissioned officers, the two companies could not well have supplied more than an average of four European artillerymen per piece *without any relief;* the deficiency being supplied by lascars, and probably also by some golandáz.

February

We now come back to the 5th Division, which, after Colonel Adams had settled matters with the remnants of the Pindáras, returned in the beginning of March to Hoshangábád.

March

4th Troop, 3rd Brig. Bengal H.A.

Captain Rodber, who had been appointed to the 6th (Native) Troop, had now taken command of it.*

* It still consisted of only four guns.

1818 March

At this time Báji Ráo, the ex-Peshwa, after his defeat at Áshti on the 20th of February, had fled to Nágpur, where Ápa Sáhib was anxiously hoping for his assistance to make head once more against the British. On the news of his approach towards that state, which was not till March, Mr. Jenkins wrote to Colonel Adams, who with his wonted energy left Hoshangábád on the 23rd, and reached Multai on the 26th, with the 5th Cavalry and two native infantry regiments; whence, Báji Ráo's movements being uncertain, he came on more slowly to Nágpur, where he arrived on the 5th of April. The ordnance train came afterwards. Lieut.-Colonel Scott had been detached towards Chánda with a small force, to prevent the Peshwa entering that town.

April

In the beginning of April, Báji Ráo was moving between Won, not far from the right bank of the river Wardah, and Idalábád, south of the Pain-Ganga. Generals Doveton and Smith, with the 2nd and 4th Divisions, had agreed to advance and prevent his escape, either northward into Málwáh, or southward across the Godáveri.

The former left Jálnah on the 31st of March, reaching Maiker on the 6th, Kárinja on the 11th, and Pandar Kaorah on the 17th of April.

The latter was at Kárlah, a few miles from Jálnah, on the 27th of March, and moved by Parlur, Sailu, Parboni, Nander, to Mudal and Belhalli, where he halted for intelligence on the 19th to 22nd of April. His march was slow, owing to the condition of his cattle.

Colonel Adams, meanwhile, whose information was always good, and whose hirkáras, well paid * and well

* Colonel Adams's head jemadár of hirkáras was a man named Rám Singh. Captain Macnaghten, in a biographical sketch of the colonel,[1] says of him:—"There was no camp and no fort into which that man

[1] *East India United Service Journal*, vol. x. p. 382.

drilled, were constantly backwards and forwards between the enemy's camp and his, left Nágpur and marched by Gumgáon and Sindhi to Hinganghát, where he arrived on the 9th. He was joined by Lieut.-Colonel Scott, and his force then consisted of only the 5th and 6th Bengal, a squadron 8th Madras Cavalry, 1st Rohilla Horse, and some native infantry, with two Madras Horse Artillery guns under Lieutenant N. Hunter, "a very distinguished officer." A despatch was sent to summon Captain Rodber, with his troop, from Nágpur. The order was received on the evening of the 15th, and the march made by Captain Rodber being one of the longest on record* made almost at a stretch, it will be best related in his own words:[1]—

1818 April

"I could not have gone less than ninety miles. At 1 a.m. on the 15th of April I commenced my march (after going the previous day eighteen miles over an execrable road—hard loose stones and hills) to join Colonel Adams at Hinganghát; on the road I received a letter, informing me that Colonel Adams had marched and encamped at Álamdoh. This information obliged me to retrace my road for some distance. I reached the colonel's camp between two and three p.m. (I am not quite positive to the time exactly; I could not have gone less than fifty-six miles, as I pushed on). At 8 p.m. same day I mounted again, and marched with the detachment in pursuit of the Peshwa, and was not dismissed till between twelve and one next day. About six of the last miles were over a succession of wooded heights, covered with large loose stones, and hollows deep and wide enough to hide both gun and horses in."

Colonel Adams, in camp at Álamdoh, on the 16th, heard that the Peshwa was at Pipalkot, 18 miles off.

would not contrive to enter undetected. . . . . Gifts and medals were conferred on Rám Singh with no niggard hand, but the Home Government would not consent to pension him, for fear of the precedent."

* "The length of this march has been variously stated. The surveyor-general of India, Colonel Thuillier, has therefore kindly verified the distance from Sitabaldi to Siuni for me. It is 95 miles, not including the length of road retraced. The route is given in the appendix to this chapter, Note G.

[1] "Letter from Lieut.-Colonel Rodber to Captain Napier Campbell, dated Kurnal, September 17th, 1832.

1818 April

This was not quite correct; he had been there, but was now at Siuni, where he heard of General Doveton's arrival at Ghát Ánji, only a few miles off, on his way towards Pandarkáorah. Of Colonel Adams's position he was not so well informed. That officer prudently took the precaution of surrounding his camp with videttes from the Rohilla Horse, to prevent any one from leaving it. At 9 o'clock the gong of a single regiment in camp sounded the hour, and forthwith the whole turned out, prepared to march. On reaching Pipalkot, after a short halt at daylight on the 17th, the colonel pushed on with the light infantry battalion, the horse artillery, and the cavalry, leaving the rest to come after. They were not much more than five miles from Siuni when the colonel, riding in front with the advance, came suddenly upon the van of the Peshwa's army, retiring from its ground at that place. He had only time to gallop back to the light infantry, which threw itself into square, and call up the horse artillery and cavalry, which came up to the front and soon got into action. Notwithstanding the great disparity in numbers (for his whole force amounted to about 1200 men, while the Peshwa's was not far short of 20,000), the actual fighting was all upon one side. The first gun that opened upon them was one of Lieutenant Hunter's, which killed the enemy's *beniwálá* (quarter-master general.) They fled from before the grape and shrapnel, and the cavalry charged whenever the ground admitted of it; Colonel Adams, at the head of the 5th Cavalry, leading the charges. The other regiment, whether from pique on the part of the commanding officer,* or from some other cause, forgetful of the cha-

* This was Lieut.-Colonel Robert Gahan, who had been so favourably reported on for previous service in this campaign, that he was gazetted a Companion of the Bath on the 18th October, 1818. He died at Jubbulpore, the 12th of December following.

racter earned for them by the gallant Fitzgerald at Sitabaldi, allowed the 5th to reap their share of a glorious success. For five miles the pursuit was kept up over strong and bushy ground. One of the Madras guns had the misfortune to get jammed in the stump of a tree, and Lieutenant G. R. Crawfurd went on with the other one. At length, as this officer said, they were so dead beat they could do no more. Two of the horses dropped dead from mere fatigue, and the rest could scarce drag the guns into the ground they encamped on. Two men wounded formed all the casualties on the British side, but there were upwards of a thousand bodies of the enemy found; and Captain Rodber, with Lieutenant Hunter, had the satisfaction of knowing that to them the principal share of credit was awarded. Verily their sleep that night must have been sweet. All accounts corroborate the following passage in Colonel Blacker's memoir:—

1818 April

"Great praise has been given to the horse artillery for their service on this occasion; and from a comparison of several accounts of this affair, whatever loss was sustained by the enemy is chiefly attributable to their fire. The nature of the country was certainly unfavourable for the charge of cavalry; yet the guns, by admirable exertion, were advanced, and the cavalry may be said to have only covered them."

General Doveton, though he discovered the enemy only from the accounts he received at Pandarkáorah of Adams's success, was able to take up the pursuit and press it hotly. Báji Ráo's force was broken up, never to be reunited. Colonel Adams therefore was at leisure to turn his face towards Chándá, which still remained to be taken. After having received, at Hinganghát, two 18-pounders from Hoshangábád, and some Madras troops under Major Goreham, Madras Artillery, who brought a third 18-pounder gun with him, he marched for that

May

1818 May.

place on the 5th, and appeared before it on the 9th of May, with the following force :—

6th Troop Bengal Horse Artillery (four guns), under Captain Rodber.
— Troop Madras Horse Artillery (two guns), under Lieutenant Hunter.
5th Company, 2nd Battalion Bengal Artillery, under Captain McDowell.
— Company, — Battalion Madras Artillery.
Two companies Pioneers (Bengal and Madras).
5th Regiment Bengal Cavalry.
6th ,, ,, ,,
One squadron 8th.
1000 Nizam's Horse.
1—19th and 1—23rd Bengal N.I.
1—1st and 1—11th Madras N.I.
Nine flank companies Bengal N.I.
Four ,, ,, Madras N.I.

Major Goreham commanded the artillery, and Lieutenant Walcott was commissary of ordnance. Lieutenant Alexander Anderson (Madras) was the senior engineer, and Lieutenant G. R. Crawfurd, Bengal Artillery, his assistant. Beside the field guns, the following pieces were all that were in park:—Three 18-pounder guns; four 12-pounder and four 6-pounder bronze guns; and six 5½-inch howitzers.

Plate No. XL.

Chándá, a large town with a circumference of five and a half miles, is situated at the juncture of the Virái and Jharpat, two small streams, dry in the hot season, which border it on the west and east. It is surrounded by a stone wall with a crenellated parapet and broad rampart, still in perfect preservation.[1] To the north lay a large piece of water, at the foot of wooded hills; and on the left bank of the Jharpat, to the east and south-east, lay Lálpettah and Bábupettah, two extensive suburbs. The tracing of the wall was broken by many re-entering angles, making enfilade available only for short distances.

[1] "Central Provinces Gazetteer, 1870." Art. "Chánda."

n
it
n

l-
it
r,
is
es
s;
x

re
ši
n,
d
d
y
n
t,
s.
g
s.

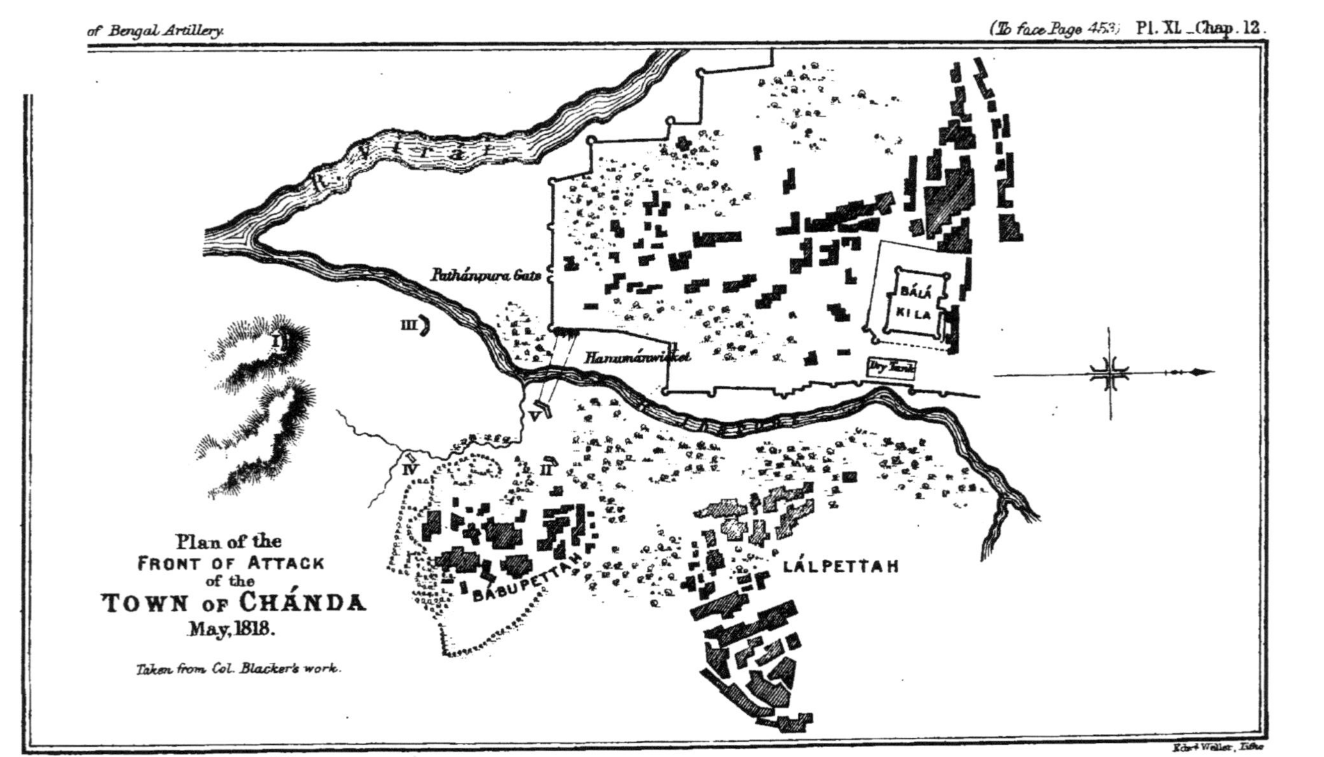

London: Henry S. King & Co. 65, Cornhill.

Colonel Adams at first encamped at Kosárá, to the north-west; but after a reconnaissance, it was determined to attack the south-east angle, as the groves about the suburbs and a ravine running down to the Jharpat afforded good cover close up to the banks of the latter. On the 13th, therefore, the camp moved to a position two miles south of the city, and Captain Rodber, with his battery and a troop of cavalry, occupied Bábupettah. The same day a sunken battery (No. I.) for two guns and a position for a howitzer were commenced upon, in order to silence some large guns on the south face which were troublesome. The pieces opened at daybreak on the 15th, and had the desired effect.

1818 May

On the 17th, the requisite materials having been prepared, the following batteries were at 8 p.m. commenced upon:—

No. II., for four 12-pounders, at 400 yards, to fire on the parapet to the right of the south-east angle, where the breach was to be.

No. III.—A sunken battery for three 6 pounders, to enfilade the same face, at the same distance.

No. IV., for two howitzers, at 630 yards.

May 18*th.*—The batteries opened at daybreak, and as the 12-pounder guns made but little impression on the wall, two 18-pounders were taken down to No. II., where they were used with effect. It was found unnecessary to construct a trench of communication to No. I., as the enemy's fire was kept down sufficiently. In the evening, the Bengal Company of Pioneers, assisted by 100 sepoys, commenced the breaching battery (No. V.) for three 18-pounders, at 200 yards. They were relieved at midnight by the Madras Pioneers, who finished it.

May 19*th.*—No. V. battery opened at 7 a.m., and continued at work during the day, interrupted frequently, however, to cool the heated metal of the guns. The shoulder of this battery was enlarged, for a 12-pounder

1818 May

to play on the defences flanking the south-east angle. At 4 p.m., a good breach one hundred feet wide was effected, but the distance from camp being too great for immediate assault, the storming was ordered to take place next morning.

May 20th.—The 6th Cavalry and Nizam's Horse had been distributed round the town, to meet the fugitives when the storming party advanced, under the general command of Lieut.-Colonel Hopeton Scott. It was in two columns, commanded respectively by Lieut.-Colonel G. M. Popham, Bengal N.I., and Captain C. Brooke, Madras N.I., with a reserve under Major R. Clarke, 5th Bengal Cavalry, which consisted of the Bengal Light Infantry, some dismounted troops of the 5th Bengal Cavalry, and two horse artillery guns, under Lieutenants Poggenpohl and Hunter. The advanced sections of the stormers were accompanied by a detail of artillerymen provided with sponges and spikes for whatever contingency might present itself. Lieut.-Fireworker Twemlow accompanied these. Every gun which could be brought to bear upon the defences kept up an active fire to the last moment, and the advancing columns met with but little opposition till they got inside. By 7 a.m. all the fighting was over.

The facilities of the ground enabled the senior engineer to dispense with many of the ordinary operations of a siege; and the great extent of the place, with a garrison of only between two and three thousand men, and its straggling buildings inside, made success more easy of attainment. But nevertheless, the judgment shown in the placing of the batteries was sound. Thus it was that a good breach was made in a few hours in a strong stone wall with only three 18-pounder guns,—and the effective working of the batteries under the deadly rays of an Indian summer day entitle the artillery to high praise. Their casualties in this siege were: Bengal Artillery—3 gunners killed; 4 gun and 2 magazine lascars, 2 drivers wounded. Madras—1 field officer and one driver killed; 1 gunner wounded. The officer was Major Goreham, who died on the evening of the 20th, from the effects of fatigue and exposure to the sun. In him the Madras Artillery lost a good soldier.

1818 May

The general order publishing the intelligence of this siege said:—

"The rapid demolition of the enemy's defences, and the speed with which a breach was effected, would sufficiently testify the science of Lieutenant Anderson, field-engineer, and of Lieutenant Crawfurd, of the Bengal Artillery, acting as engineer, in indicating the positions for the batteries, even had not Lieut.-Colonel Adams professed his obligations to those officers so warmly.

"It is distressing that Major Goreham has not survived to enjoy the just reputation which his eminent merit in the command of the artillery challenged for him; yet, if he sunk under his too earnest exertions, he bore with him to the tomb the universal admiration of the army, and his name will long be quoted to excite similar energy in others. Captain Rodber, Captain McDowell, Captain Mackintosh, and Lieutenant Walcott, seem to have highly deserved the praise which their commander bestows upon them. Indeed, the efforts of all the officers and men of the artillery appear to have been proudly laudable; and in particular the successful attempt of Lieutenants Poggenpohl and Hunter to get one of the guns of the Horse Artillery over the breach, exhibits a spirit and a resource of superior tone."[1]

Besides the officers of this branch already mentioned, Lieutenant P. Montgomerie, of the Madras Regiment, was present. There appear to have been but few, on the whole, and the number available was still further reduced by the appointment, in June, of Lieutenant Gavin R. Crawfurd, who had been acting as engineer to the office of superintendent of the Chándá district, which he held till February, 1825. The ability with which he discharged

[1] G. G. O. Gorakhpore, 18th June, 1818.

1818 its duties has been placed upon record by one of his successors, Major Lucie Smith.[1]

1817 No notice has hitherto been taken of the Reserve Division, under Sir David Ochterlony. His duties were political rather than military. In December he moved towards Jeypore, where the half-Pindára Amir Khán was. A treaty was concluded with this chieftain, in which he consented to disband his army and give up his guns—terms that he carried out as far as he was able.

1818 But Jamshed Khán, one of his generals, encamped near the Sámbar lake with a body of troops and 44 guns, did not comply. Brigadier A. Knox, therefore, was

April detached in the beginning of April, with the cavalry brigade, three battalions of native infantry, and ten guns, to take them. Coming up, after a few forced marches, with the Pathán, in a few spirited words, as Colonel Neville Parker had done before at Kora, he gave him his option of yielding or proving his ability to keep the guns. Jamshed Khán preferred discretion to valour.

At the same time, Brigadier J. Arnold, who had been left in command at Rewári, was sent into the districts of Hariána and Bhatiána, to reduce the forts of Fathiá-

B—C R.H.A. bád, Susa, and Rániya. The 3rd Troop, commanded by Captain J. H. Brooke, with half the rocket troop, marched from Meerut on the 23rd of August, under command of Major Stark, to join this force. The 4th Company, 1st Battalion, under Captain C. H. Bell, joined it from Kurnal, taking up a small battering train at Delhi on the way.

Following out the same policy of restoring their own to the chieftains of Rájputána, the towns and forts in

[1] "Central Provinces Gazeteer." Art. "Chándá."

the various states, which had either been seized by others or were in a state of rebellion, were successively reduced. Táragarh, the citadel of Ajmir, garrisoned by some of Sindiah's troops, was in June compelled to surrender to Brigadier Knox, after batteries had been erected. Major E. W. Butler commanded the artillery. Lieutenant C. Smith,* who had joined from the Centre Division, was also present. Ensign E. Garstin was the engineer officer. Mádhurájpura † was taken by a small force under command of Lieut.-Colonel W. A. Thompson, C.B., Bengal N.I. Ground was broken before this place on the 27th of July; and on the evening of the next day, the breach in the wall of the town being reported practicable, three columns were ordered for the storm—one by the breach, two by escalade. All were successful; and on the 1st of August the fort within the town capitulated. The casualties were—one gunner killed; a lascar and an ordnance driver wounded. The artillery officers ‡ are thus mentioned in Lieut.-Colonel Thompson's report:—

1818 June

July

August

"Major Butler and the officers and men of the artillery sustained the well-known reputation of that distinguished corps, and are entitled to the highest commendation. The practice of the mortar and breaching batteries (the former under command of Captain Pereira, and the latter under Captain Graham, who zealously volunteered his services to command the breaching batteries, both against the town and fort) was most excellent; to which, and the fire from the other batteries under Lieutenants Smith, Baker, and Whinfield, the surrender of the fort with so small a loss on our side must be attributed."

* Captain Graham, the adjutant, must also have been present, and probably also the officers of both the 4th and 5th Companies, 1st Battalion.

† About 26 miles south-south-west from Jaipur.

‡ Only the 5th Company, Captain Pereira's, appears to have been present here. Lieutenant Oliphant, of the 5th Company, may have been doing duty with the 4th, which was short of officers, but his name does not occur in the order.

1818 October

The fort of Nasridah * was also taken in the following October. Lieutenant Smith was here, and, as he does not appear to have been senior, perhaps also Major Butler. Other minor operations were undertaken in Rájputána during the next four or five years, but they do not require mention here.

1819

The last siege in this campaign in which the Bengal Artillery took a part was that of Asirgárh, ceded by Sindiah in the treaty of the 5th of November, 1817, but still held by the Kiladár, Jaswant Ráo Lár, who it was known would not give it up peaceably. But it had not before been found convenient to attack it in force. Oblong in shape, about 1100 yards by 600 in extent, and situated on the level summit of a rock 750 feet above the plain, its sides everywhere precipitous gave it a character of almost impregnability, though it was not a maiden fortress, having been taken by the Emperor Akbar in A.D. 1559, and again by a detachment from Major-General Wellesley's army in 1803.

General Doveton was encamped in the latter end of February, 1819, at Kálá Chabutra, close to Asirgárh, on the road to Burhánpur, and Sir John Malcolm had moved across the Nerbudda to Sandalpur, on the northern side.

March

On the 1st of March a detachment arrived in General Doveton's camp, escorting a battering train from Jálnah, and on the 11th a further reinforcement from Nágpur

E-20 R.A.

joined. The C Company of the 1st Battalion Madras Artillery, with a company of lascars, which then formed a part of the Nágpur subsidiary force,† was probably the one which accompanied this part of the army.

* Situated on the north of the Banás river, south-east from Tonk Rámpura, and about 75° 23′ east longitude, and 25° 58′ north latitude.

† Madras Artillery Records. Lieutenant G. Twemlow, who was acting adjutant to the force, had volunteered for this service, and joined about this time.

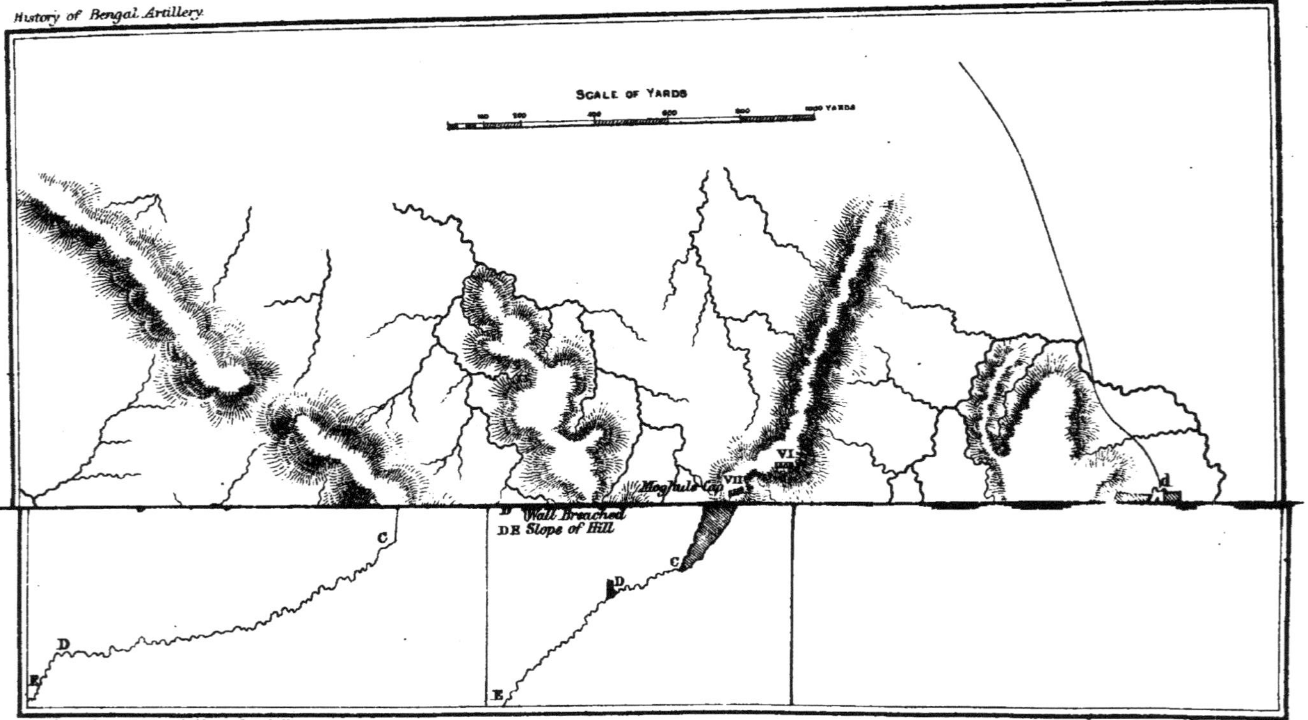

SCALE OF YARDS

100 200 400 600 800 1000 YARDS

Moghuls Gap

VII

VI

d

C

D

E

D Wall Breached

DE Slope of Hill

London: Henry S. King & C°. 65, Cornhill.

Edwd. Waller, Litho.

Another detail of heavy guns from the Hoshangábád battering train, with a detachment of native infantry, under Lieut.-Colonel J. Greenstreet, arrived in camp on the 17th. Lieutenant J. E. Debrett, with the head-quarters and half the 5th Company, 2nd Battalion Bengal Artillery, accompanied it. The two divisions now were constituted as follows:—

1819 March

4–23 R.A.

BRIGADIER-GENERAL DOVETON'S DIVISION.

Artillery.—A Troop Madras Horse Artillery; half of the 5th Company, 2nd Battalion Bengal Foot Artillery.

A—D R.H.A.

Pioneers.—300 Bengal Pioneers; Detachment Madras Pioneers.

Cavalry.—6th Bengal Cavalry; 2nd and 7th Madras Cavalry.

European Infantry.—Royal Scots (1st Royals); one wing 30th Regiment; 67th Regiment; Madras European Regiment.

Native Infantry.—1—15th and 2—15th Bengal N.I.; 1—7th, 1—12th, 2—17th, 2—13th, and 2—14th Madras N.I.

BRIGADIER-GENERAL SIR JOHN MALCOLM'S DIVISION.

Artillery.—Half of the B Troop Madras Horse Artillery; D Company, 2nd Battalion Madras Artillery, with camel howitzer battery under Captain Frith.

B—D R.H.A.
2–17 R.A.

Pioneers.—Detachment Madras Pioneers; Detachment Bombay Pioneers.

Cavalry.—3rd Madras Cavalry.

Infantry.—2—6th and 1—14th Madras N.I.; 1st Battalion Grenadier, and 1—8th Bombay N.I.

The artillery was commanded by Lieutenant J. Crosdill, C.B., who had been nominated to the distinction of the Bath, then a rare one in the Indian army, for his previous services in this war. His brigade-major was Captain Poignand; and Major Weldon was the commissary of stores. The ordnance having been received at different periods before and during the siege, a detail of it is given in the appendix to this chapter (Note H).

The engineer department was commanded by Lieutenant John Coventry, Madras Engineers, with the following officers:—Lieutenants John Cheape and A.

1819 March

Irvine, and Ensign T. Warlow, Bengal; and Lieutenant John Purton and Ensign E. Lake, Madras establishment. Their working strength consisted of—

Madras Sappers and Miners—European and native ...	85
Bengal Miners—native	25
Bengal, Madras, and Bombay Pioneers—native ...	1000

The ramparts of this fort were of masonry, thick and lofty, with cavaliers at intervals, mounted with large guns, which commanded the country on every side. One of these, from a cavalier on the south-west, after lying there uncared for, for fifty years, was not long ago ordered to be sent home to Woolwich.* On the westward, for nearly half the circumference, a second rampart follows the outline of the wall, enclosing a narrow space called the "Kamargáh;" and lower down on the same side lay the "Máligarh," or lower fort, beyond which was the town, in a hollow intersected by ravines, and everywhere commanded by the works of the lower fort. The principal entrance to the upper portion is at the western extremity. It was secured by five gateways of good masonry, and steep flights of stone steps led down from it into the lower fort.

The natural defence of the fort fails in three places. One (marked *a* in Plate XLI.) on the north side, where a thick double rampart supplied the deficiency. The second (marked *b*) in a ravine near the north-east angle, running into the fort, across which a casemated rampart nearly fifty yards in length and forty feet thick was built. The third, near the south-east angle (at *c*), where was a sally-port open at top, and protected by five traverses, with a wall in front to prevent the entrance being seen from the country.

* A description of this gun is given in the appendix to this chapter (Note K).

1819 March

For some time previous to the commencement of the siege, a large depôt of materials had been forming; and on the 17th orders were given for the assault, by both divisions, of the *pettah*, or town, in the buildings of which cover would be found for the intended attack on the lower fort.

March 18*th*.—The town was occupied to-day with little loss, notwithstanding a continued fire from the fort, by columns from both divisions. The engineer depôt was established in a bomb-proof pagoda in the centre of the town. A battery (No. I.) for six light howitzers was completed during the day in the same place, to keep down the enemy's fire. The troops occupied the street (*d d*) running parallel to the fort. During the night a breaching battery (No. II.) for six 18-pounders and two 12-pounder guns for the north-west angle of lower fort was begun, but owing to the hardness of the ground and want of materials, it was not continued for the present. Those streets running in the direction of the fort were barricaded at their entrance, to secure them from attack* as well as from enfilade. A good post (*e*) was also completed at the salient point of the town; but as its communications were flanked from the lower fort, it was only used as an outpost at night, the men in it being withdrawn by day to *f*.

March 19*th*.—The enemy made a sally at sunset, and succeeded in burning a few houses about the outpost (*e*). During the night battery No. II. was completed, and battery No. III., on the north side of the town, close to the pagoda, was begun; but the ground was so hard that the pickets fastening the fascines (the only material at hand) of which the revêtment was made could not hold with sufficient firmness, and the work had to be postponed.

* Not effectually, as the subsequent attack by the enemy proved.

1819 March

March 20th.—The guns opened at daybreak with good effect, silencing the enemy's fire, and by the evening making a practicable breach in the works of the lower fort. At this time the enemy, making a bold sally, effected an entrance into the main street of the town. They were repulsed; not, however, without the loss of Lieutenant-Colonel Fraser, of the Royal Scots, who was killed.

March 21st.—An unfortunate accident occurred in No. II. battery at seven this morning. The expense magazine containing 130 barrels of powder* exploded, killing or wounding 2 native officers and 83 rank and file out of 100 sepoys of the 2—15th Bengal N.I., who were on duty in the battery.† On this the enemy, who had shortly before, it was afterwards found, evacuated the lower fort, returned again and opened fire; but the battery continuing to work without intermission, it soon silenced theirs. Battery No. III., completed the day before with sand-bags, opened to-day at 3 p.m.

March 22nd.—No. III. battery sent 130 shells into the place to-day. Two 12-pounder guns were placed,

* There appears to have been an unnecessarily large quantity of powder for an expense magazine, exposed as it must be to many chances of accident in the hurry of action as well as from the enemy's fire. Thirty barrels of powder L.G. would have sufficed for one gun per minute throughout the day, allowing time for the guns to cool; forty would have been an ample allowance for two days' expenditure. The position of the magazine was good. It was placed against the perpendicular bank of a deep ravine, thirty yards to the left rear of the battery, and probably the native infantry availed themselves of the shelter of this ravine; but the sepoys must have been very close to the magazine, and natives are proverbially careless regarding powder. The officer in command of them, Lieutenant Malcolm Nicholson, escaped, probably being in the battery at the time.

† Lake and Blacker put down the casualties at 99 rank and file; but the above, from a record of the services of the 15th N.I. in the *East India United Service Journal* (vol. v. p. 72), seems more probable. Blacker's figures appear to comprise the casualties of the regiment during the siege.—See his Appendix A A.

one 200 yards to the right, to destroy some defences of the lower fort; the other the same distance to the left, to silence a large gun in the north centre bastion which was troublesome.

1819 March

March 24th.—The following is an extract from Lieutenant Coventry's report to the general, after a reconnaissance made by him the day before, of the east front of the fort, which was to be attacked :—

"The irregular nature of the ground, and the cover afforded by ravines, render extensive parallels unnecessary. A communication, however, should be opened from the Rám Bágh to a ravine on the left of the attack, to enable the working parties to arrive under cover.

"As the approaches are to be carried up a ravine, exposed to a direct fire in front and a flanking fire on each side, it becomes an object of the first importance to knock off the defences of the flanks, and prevent the enemy from rolling down stones. I recommend that these works should be destroyed from the foundations.

"This, I conceive, can be effected by placing batteries on the prolongation of flanks, in such manner as will enable us to breach the opposite, and enfilade the adjacent flank from the same battery. The flanks being destroyed, and the defences of the curtain wall knocked off, the bottom of the revêtment of the retaining and curtain walls is to be loosened, to enable the miners to establish themselves; or, should this be found impracticable, a breaching battery to be constructed, and the curtain wall laid open. I am of opinion that by one or other of these means we shall be able to form a practicable breach.

"The mortar batteries to be disposed as represented in the plan, and, if practicable, a brigade of 6-pounders to be placed in battery on the detached hill opposite south-east angle, so as to command the high ground in rear of the front attacked.

"To distract the enemy's attention from the real point of attack, it is advisable that, the evening previous to constructing the batteries, possession should be taken of the lower works on the *pettah* side, and a battery constructed to play upon the gateways.

"By these means the garrison will be deprived of all hopes of escape, and their uncertainty as to the true point of attack will weaken their efforts to oppose us.

"On the same principle, I recommend that the south-east face should be breached where the rock fails, with a view to such advantage being taken of it as circumstances may require."

1819 March

The point *a* was however selected instead of the south-east face, the attack of which would have distributed the batteries over too wide a space, and in fact would have constituted three separate attacks instead of two—the east and west, as was now determined on. A battery (No. IV.) for two 8-inch howitzers and two 5½-inch mortars was placed 350 yards to the left front of No. II. The enemy kept up a smart fire from the lower fort during the night.

March 25th and 26th.—Destroying the defences on either flank of the breach in the lower fort, and bombarding the upper fort.

March 27th.—East Attack.—The engineer's depôt having been established at the Rám Bágh, the enemy brought a large gun in the north-east bastion (*g*) to bear upon it, by which many carts were destroyed; so a battery (No. V.) was thrown up in front of the garden for two 12-pounders.

West Attack.—No. VI., for an 18-pounder and a 12-pounder gun for making a breach in the lower fort, was commenced this day in a hill to the south-south-east, and the guns for it were brought up by elephants to the eminence on the same ridge, called "The Mogul's Cap" (No. VII.).

March 28th.—Completing communications to batteries No. VIII. and IX. on the east side, which were not yet begun. On the west the 18-pounder gun was placed in No. VI.; the 12-pounder broke down.*

March 29th.—East Attack.—No. VIII., for five 18-pounder guns, was constructed during the night, 380 yards from the north-east angle, and 530 from the end of the opposite flank (*b h*), the defences of which it was

* It was replaced, though Lake does not say so; but he uses the word "guns" in speaking of this battery. Blacker (G, p. 419) calls it "the four-gun breaching battery," which is probably correct.

1819 March

intended to destroy. No. IX., for four 18-pounder guns, was prepared at 350 yards from the fort at *h*, and 600 from the flank *b g*.

West Attack.—Nos. II. and VI. opened at daylight, one to improve the old breach, the other to make a new one. Both were reported practicable in the evening, and the assault of Máligarh was ordered for next morning. A $4\frac{2}{5}$-inch howitzer was placed to the right of No. VI., to command the gateway of the upper fort.

March 30th.—East Attack.—Conveying guns up the heights—heavy work even for a battalion of Europeans and one of natives, assisted by elephants. Only three were got into No. VIII., on the left of which a place for two heavy mortars was made.

West Attack.—The enemy evacuated Máligarh this morning, and it was immediately occupied. No. II. battery was dismantled. All the mortars but one were removed from the town into the lower fort.

March 31st.—A reinforcement from the troops at Ságar of 2200 men arrived to-day. Brigadier-General Watson, C.B., commanding it, had joined the day before. The 6th Company, 2nd Battalion, and a company of golandáz, Bengal Artillery * were with it. The battering train was not all in before the 3rd of April.

Battery No. X., for eight mortars and howitzers, was constructed under the hill to the right front of the Rám Bágh. A 12-pounder was added to the right of No. VIII., to keep down the matchlock fire from about *g*. No. XI. battery, for mortars and howitzers, was completed and armed, and two $5\frac{1}{2}$-inch howitzers were sent to the Mogul's Cap, No. VII.

* And the following officers:—Lieutenants G. R. Scott, W. E. J. Counsell, and T. D'Oyly. The first-named officer was adjutant to the Bengal Artillery. Captain Pew, commanding this company, was on duty at Ságar.

1819 April

April 1st.—No. X. opened to-day. The embrasures of No. VIII. were repaired and widened. A ten-mortar battery, No. XII., was thrown up on the left rear of No. IX., and armed next day.

In the west attack a six-gun battery, No. XIII., was constructed to breach the wall of the Kamargáh; it opened next day. An 18-pounder and a 12-pounder gun were also placed in the town, to enfilade and destroy the defences of the same work.

April 2nd.—Batteries firing with good effect.

April 3rd.—East Attack.—No. XII. battery opened this day. A four-mortar battery was thrown up about 100 yards to the right front of No. VIII.

West Attack.—The defences to the right of the intended breach having been destroyed, a parapet was thrown up on the ridge above No. XIII., where a breaching battery further advanced was to be placed.

April 4th.—East Attack.—The flanking defences of the casemated rampart being nearly destroyed, an advanced breaching battery for two 24-pounder and four 18-pounder guns (No. XIV.) was commenced. As there was not sufficient space for all, the guns were placed on three different levels in echellon. Three 18-pounder guns were added to the right of No. XIII., to bear upon the bastion *g*, the fire from which was very annoying.

West Attack.—A mine under the rampart of the lower fort was commenced, to open a road for the guns into No. XV. battery.

April 5th.—East Attack.—Battery No. XIV. was completed. It was 400 yards from the retaining wall below the casemated battery, and 450 from the latter; a magazine for it and a road for the guns was also made. Another (No. XVI.) was constructed in front of No. IX., to breach the retaining wall. To-day a 140-pounder gun on the north-east bastion (*g*) was brought down and rolled half-way to the bottom of the hill.

1819 April

West Attack.—The breach in the Kamargáh being completed, No. XV. was commenced. The guns for it—four 18-pounders—were dragged up into the lower fort. The mine was sprung and a good road opened for the guns.

April 6th.—Batteries Nos. XIV. and XV., in the east and west, were armed to-day, and No. XII. was repaired.

April 7th.—It appeared to-day that the Kiladár had begun to look on his case as hopeless. The fire on both points of attack being very effective, the garrison became alarmed, and proposed terms of surrender, which however were not accepted.

April 8th.—The breaching batteries reopened at daylight, but at 11 o'clock a.m. were ordered to cease firing—Jaswant Ráo Lár had surrendered unconditionally.

Next day at sunrise the garrison marched out, and delivered up their arms. The union-jack was hoisted under a royal salute from all the batteries. Owing to the good cover, the enemy only lost 43 killed and 95 wounded; of the former, his jemadár of artillery was the most felt as a loss by Jaswant Ráo, but the failure of his ammunition * must have compelled a surrender. The casualties of the artillery are given in the appendix, Note I.

This was the last act of the Máhrátá war. There had not been left, at the close of 1818, a single spot in Central India which a Pindára could call his home,[1] and now English troops garrisoned the last of the Máhrátá strongholds which had resisted our arms. One by one

* There was but 2 cwt. of powder left in the magazines.—Prinsep, ii. 331.

[1] Sir John Malcolm.

1819 April

the great powers of that nation had fallen before the Lion and Unicorn, compelled by energy and courage directed by the genius of a good soldier and statesman. The errors of Lord Hastings in his management of the Nipál war were corrected in this one, for every officer entrusted with command showed himself fully capable. The task of hunting down an enemy inured to rapid and unexpected movements through districts, many of them still imperfectly known and mapped, required careful combinations; but the chances of opposition arising from five* great Máhrátá states and other lesser powers were so numerous, that their calculations seemed impossible; yet, as Prinsep observes, "Not one adverse circumstance or occasion of danger arose, without its remedy and corrective being found ready at hand."

The work performed by the troops was necessarily severe. The incessant marching after the scattered bodies of Pindáras and the ex-Peshwa, gave the campaign of 1817-18 a distinctive character of its own, repeated in our own times in the pursuit of Tántiyá Topi. In these rapid movements the horse artillery had their full share; the dismounted branch, as it then almost entirely was, less so, being chiefly attached to the battering trains. The former, of the Bombay corps, admirably horsed as it always has been, which belonged to the 4th Division under General Lionel Smith, covered 2250 miles in seven months—"a movement," says Colonel Blacker, whose statements are not rashly made, "of which there is probably no example in the world." The same division, with its battering train (not a large one), marched 300 miles in twenty-six days; and General Pritzler's force, with field-guns, 346 miles in twenty-five

* The Gáikwár—head of the Baroda state—was the only one who did not require direct coercion.

1819 April

days.[1] Sir William Keir's column for two months was scarcely ever at rest, till the confines of Málwá were cleared. Equally energetic were Generals Doveton, Marshall, and Brown, and Colonel Adams, whose operations have been noticed already.

Owing to the deficiency of artillery and engineer officers their regimental duties were very heavy, and at the siege of Asirgárh the former actually lived in the batteries to which they were detailed. Fortunately, throughout the war but few casualties occurred in the artillery—but one of them, Major Goreham, was the result of over-fatigue under a summer sun.*

There was also a great want of siege ordnance in the Madras and Bombay divisions. The siege of Maligám was commenced with two 12-pounder and two 18-pounder guns, which were in action from the 19th till the 27th and 29th of May respectively, by which time 5800 rounds had been expended, and the dimensions, *externally*, of the vents were as follows :—

18-pounder gun	...	...	$2\frac{1}{4}''$ by $3''$
,,	...	...	$4\frac{1}{3}''$,, $3\frac{3}{4}''$
12-pounder gun	...	...	$1\frac{1}{3}''$,, $1\frac{1}{8}''$
,,	...	...	$2''$,, $2''$ [2]

Such a result plainly points to previous use of pieces defectively cast; but it is a well-known fact that a considerable number of the iron smooth-bores still in the forts and arsenals of India were taken from ships either paid off, or captured during the French war, and are only fit to be broken up.

* On one occasion, at the siege of Belgám, when the enemy, taking advantage of an explosion in one of the batteries, sallied out to attack it, the artillery officer, Lieutenant W. F. Lewis, formed up his gunners, and, along with the detachment of sepoys in the battery under Lieutenant Walker, charged and drove them back into the town.—Blacker, p. 293.

[1] Blacker, p. 293. [2] Papers, p. 371.

AUTHORITIES CONSULTED FOR THIS CHAPTER.

1. Memoir of the Operations of the British Army in India during the Máhrátá War of 1817, 1818, and 1819. By Lieut.-Colonel V. Blacker, C.B. 1 vol. 4to. London, 1821.

2. History of the Political and Military Transactions in India, during the administration of the Marquis of Hastings. By H. T. Prinsep, Esq. 2 vols. 8vo. London, 1825.

3. History of the Máhrátás. By Captain James Grant Duff. London, 1826.

4. Papers respecting the Máhrátá and Pindári Wars. Printed in conformity with a resolution of the proprietors of East India Stock.

5. Journals of Sieges of the Madras Army in 1817, 1818, and 1819. By Lieutenant E. Lake. 1 vol. 8vo. with atlas. London, 1825.

6. East India Military Calendar.

7. Services of the Madras Artillery. By Major Begbie. 2 vols. 8vo. Madras, 1853.

8. Copies of Muster-Rolls.

9. Madras Artillery Records.

10. East Indian United Service Journal.

APPENDIX.

Note A.—List of troops and companies of the Bengal Artillery which served in the Pindári and Máhrátá war, with the names of officers who then commanded them, and their latest designation.

Note B.—Actions and sieges in the Pindári and Máhrátá war, with names of Bengal Artillery officers who were present.

Note C.—Officers of the Bengal Artillery who served during the Pindáriá and Máhrátá war.

Note D.—Officers of the Madras Artillery who served during the Pindári and Máhrátá war.

Note E.—Officers of the Bombay Artillery who served during the Pindári and Máhrátá war.

Note F.—Officers of the Engineer Department who served during the Pindári and Máhrátá war.

Note G.—Route from Sitabaldi, near Nágpur, to Siuni, showing the distance marched by the 3rd Native Troop Bengal Horse Artillery in pursuit of the Peshwa, from April 16th to 17th, 1818, as verified in the office of the Surveyor-General of India.

Note H.—Statement of ordnance employed in the siege of Asirgárh, taken from Blacker and other accounts.

Note I.—Return of casualties in the Bengal, Madras, and Bombay Artillery, during the siege of Asirgárh.

Note K.—Description of the great Burhánpur gun, taken in the fort of Asirgárh.

Note A.

List of troops and companies of the Bengal Artillery which served in the Pindári and Máhrátá war, with the names of the officers who then commanded them, and their present or latest designation.

Latest designation.	Designation in 1817-1819.	Names of Officers in command during the war.	Remarks.
A–C R.H.A.	1st Troop	Captain J. P. Boileau	Joined the Centre Division from Meerut. Returned to Meerut in February, 1818.
A–F R.H.A.	2nd Troop	Captain (Brevet-Major) H. Stark	Joined the Centre Division from Meerut. Two guns under Lieutenant Matheson detached with General Brown's column. Served at capture of Jáwad. Returned to Meerut in February, 1818.
B–C R.H.A.	3rd Troop	Captain J. H. Brooke	Joined the Centre Division from Meerut, and from thence the Reserve in March, 1818. With Colonel Arnold's force in Bhatiyána and Bikanir. Returned to Meerut at the end of the year.
B–F R.H.A.	Rocket Troop	Captain W. S. Whish	Joined the Centre Division from Cawnpore. Proceeded to Meerut in February, 1818.
4–2 H.A.	4th Troop	Captain G. E. Gowan	Right Division. With Rájputána field force, under Colonel Ludlow. Stationed at Neemuch in 1819.
4–1 H.A.	5th Troop	Captain J. A. Biggs	Centre Division. Detached with General Brown's column. Served at capture of Jáwad. Sent to Agra in 1818.
4–3 H.A.	6th Troop	Lieutenant G. Blake	5th Division. Two guns under Lieutenant Blake at battle and siege of Nágpur. Two under Lieutenant Twemlow with Colonel Adams, following up Pindáras. The whole under Captain Rodber with Colonel Adams, in pursuit of the Peshwa, battle of Siuni, and siege of Chándá. Stationed at Hoshangábád in 1819.

NOTE A (*continued*).

Latest designation.	Designation in 1817-1819.	Names of Officers in command during the war.	Remarks.
2–23 R.A.	4th Co., 1st Bat.	Lieutenant C. H. Bell	Joined the Reserve from Kurnal in 1817. Sent with Colonel Arnold's force to Bhatiyána and Bikanir in 1818. Returned to Kurnal same year.
A–16 R.A.	5th Co., 1st Bat.	Captain I. Pereira	Joined the Reserve from Kurnal in 1817. Sent next year to Nasirábád.
Reduced 1825	2nd Co., 2nd Bat.	Captain A. Fraser	Joined the Right Division from Agra in 1817. Returned to Agra next year.
Reduced 1825	4th Co., 2nd Bat.	Captain-Lieutenant E. Pryce ... Lieutenant J. C. Carne ... ... Lieutenant T. Sanders ... ... Captain S. Coulthard	Joined the Centre Division in 1817. Detachment under Captain-Lieutenant Pryce with Colonel Philpot's force. Transferred to Left Division in February, 1818. Present at sieges of Dhámoni, Mandalah, Sátanwári (detachment under Lieutenant Carne).
4–23 R.A.	5th Co., 2nd Bat.	Captain J. McDowell ... ... Lieutenant J. E. Debrett (from Oct. 27th, 1818)	Joined the force on the Nerbudda under Colonel Adams in the end of 1816. Detachments with Colonels Adams and Macmorine in 1817-18. Present at the sieges of Chándá and Asirgárh. Proceeded to Hoshangábád in April, 1819.
4–22 R.A.	6th Co., 2nd Bat.	Captain A. Lindsay ... ... Captain P. L. Pew	Joined the Left Division from Cawnpore in 1817. Present at the sieges of Garhákota and Asirgárh.
C–19 R.A.	7th Co., 2nd Bat. (Detachment.)	Lieutenant T. D'Oyly	Only a detachment served during the war, as appears from the official notification regarding prize-money. Lieutenant D'Oyly was in command of it. It was at Jubulpore, and also at the sieges of Mandalah (April) and Garhákota (October, 1818).
B–19 R.A.	6th Co., 3rd Bat. 1st Co., 2nd Bat.	Capt. W. Battine (Brigade-Major) Lieutenant R. C. Dickson	Joined the Centre Division from Cawnpore in 1817. Returned to Cawnpore early in 1818.
Reduced 1825	7th Co., 3rd Bat. 8th Co., 2nd Bat.	Lieutenant T. Timbrell	Joined the Centre Division from Cawnpore in 1817. Returned to Cawnpore early in 1818.

NOTE B.

Actions and sieges in the Pindári and Máhrátá war, with the names of Bengal Artillery officers who were present.

Actions, etc.	Date.	Names of Officers.
Nágpur (battle)	Dec. 16, 1817	Lieut. G. Blake
Nágpur (siege)	Dec. 19—24	Lieut. G. Blake
Jubulpore (action)	Dec. 19	Lieut. T. D'Oyly
Jáwad (assault of)	Jan. 19, 1818	Capt. Biggs; Lieuts. Matheson and Kempe
Dhámoni (siege)	Mar. 19—24	Capts. Hetzler and Lindsay; Capt.-Lieut. Coulthard; Lieuts. Carne, Bell, Sanders, Crommelin, and Patch
Mandalah (siege)	April 18—26	Capts. Hetzler and Lindsay; Lieuts. Carne, Dickson, D'Oyly, Kirby, Sanders, Crommelin, and Patch
Siuni (action)	April 17	Capt. Rodber; Lieuts. Walcott and Crawfurd
Chándá (siege)	May 19—21	Capts. Rodber and McDowell; Lieuts. Walcott, Debrett, Crawfurd, and Twemlow
Sátanwári (assault of)	June 8, 9	Lieuts. Carne and Sanders
Táragarh (siege)	July 1, 2	Major Butler; Capt.-Lieut. Graham; Lieut. C. Smith. Probably also Capts. Pereira and C. H. Bell; Lieuts. Baker, Webb, Oliphant, and Whinfield
Mádhurájpura (siege)	July 27—Aug. 1	Major Butler; Capts. Pereira and Graham; Lieuts. Smith, Baker, and Whinfield
Nasridah (siege)	Oct. 25—29	Major Butler; Capt. Graham; Lieut. Smith
Garhákota (siege)	Oct. 18—30	Major Hetzler; Capt.-Lieut. Coulthard; Lieuts. Pew, Sanders, Crommelin, and Counsell
Asirgárh (siege)	March 17—April 8, 1819	Lieuts. Debrett, G. R. Scott, D'Oyly, Twemlow, and Counsell

NOTE C.

Officers of the Bengal Artillery who served in the Pindári and Máhrátá war.

Rank and Names.	Posting.	Remarks.
Lieut.-Col. G. Mason, C.B.	... ...	Commanding Artillery Centre Division
„ A. Macleod, C.B.	... ...	Commanding Artillery Right Division
„ E. W. Butler *	... ...	Commanding Artillery Reserve
„ G. Pennington *	... ...	Commanding Horse Artillery
Major R. Hetzler *	... ...	Commanding Artillery Left Division, and Ságar Field Force
„ H. Stark *	2nd Troop H.A.	
„ J. H. Brooke *	3rd Troop H.A.	
Captain J. Young	5th Co. 2nd Batt.	Military Secretary to Government. With Governor-General
„ A. Lindsay	6th Co. 2nd Batt.	
„ J. A. Biggs	5th Troop H.A.	
„ J. P. Boileau	1st Troop H.A.	
„ W. S. Whish	Rocket Troop	
„ J. Rodber	6th Troop H.A.	
„ W. Battine	6th Co. 3rd Batt.	Brigade-Major Artillery Centre Division
„ A. Fraser	2nd Co. 2nd Batt.	
„ J. McDowell	5th Co. 2nd Batt.	
„ E. Pryce *	4th Co. 2nd Batt.	
„ G. E. Gowan *	4th Troop H.A.	
„ J. Tennant *	2nd Co. 1st Batt.	
„ J. Pereira *	5th Co. 1st Batt.	
„ C. Graham *	4th Co. 1st Batt.	Acting Adjutant Artillery Reserve
„ J. Curtis *	2nd Troop H.A.	Doing duty with 4th Troop H.A.
„ T. Chadwick *	4th Co. 2nd Batt.	In charge 1st and 10th Companies Golandáz
„ J. C. Hyde *	3rd Troop H.A.	
„ S. Coulthard *	5th Co. 1st Batt.	Afterwards 4th Co. 2nd Batt.
„ P. L. Pew *	6th Co. 2nd Batt.	
„ C. H. Bell *	4th Co. 1st Batt.	
Lieut. W. G. Walcott	2nd Co. 3rd Batt.	Commissary of Stores Nágpur Subsidiary Force
„ J. C. Carne	4th Co. 2nd Batt.	
„ J. E. Debrett	5th Co. 2nd Batt.	
„ G. N. C. Campbell	Rocket Troop	Afterwards 3rd Troop

* These officers were promoted to the rank here given them by the augmentation dated 1st September, 1818. The postings in all cases are those at the opening of the war.

NOTE C (*continued*).

Rank and Names.	Posting.	Remarks.
Lieut. D. Macalister	3rd Troop H.A.	Afterwards 1st Troop
„ J. J. Farrington	5th Troop H.A.	
„ G. Brooke	Rocket Troop	
„ T. Lumsden	2nd Troop H.A.	Quarter-Master H.A. Centre Division
„ T. Croxton	2nd Co. 2nd Batt.	Adjutant Artillery, Right Division
„ P. G. Matheson	2nd Troop H.A.	
„ T. Timbrell	7th Co. 3rd Batt.	Adjutant Field Artillery, Centre Division
„ J. Sconce	3rd Troop H.A.	Adjutant Horse Artillery, Centre Division
„ G. Blake	6th Troop H.A.	Afterwards 5th Co. 2nd Batt.
„ Roderick Roberts	2nd Troop H.A.	Afterwards 1st Troop
„ C. Smith	6th Co. 1st Batt.	Doing duty with 7th Co. 3rd Batt. Afterwards with Reserve Division
„ H. C. Baker	5th Co. 1st Batt.	
„ G. H. Woodrooffe *	1st Co. 1st Batt.	(?)
„ W. Bell	7th Co. 3rd Batt.	Afterwards 4th Co. 2nd Batt.
„ H. Webb	4th Co. 1st Batt.	
„ W. Oliphant	5th Co. 1st Batt.	
„ H. J. Wood	7th Co. 3rd Batt.	Quarter-Master Field Artillery, Centre Division
„ E. P. Gowan	1st Troop H.A.	Transferred to 3rd Troop in Dec., 1817, and afterwards 4th Troop
„ F. S. Sotheby	1st Co. 3rd Batt.	Commanding Golandáz Russell's Brigade (Nizám's contingent)
„ R. C. Dickson	6th Co. 3rd Batt.	Afterwards 2nd Co. 2nd Batt.
„ E. Huthwaite	7th Co. 2nd Batt.	Doing duty with 7th Co. 3rd Batt.
„ G. R. Crawfurd	5th Co. 2nd Batt.	Adjutant Nágpur Subsidiary Force. In civil charge of Chándá district from June, 1818
„ H. Delafosse	6th Co. 3rd Batt.	
„ G. Robertson Scott	2nd Co. 2nd Batt.	Afterwards 3rd Co. 2nd Batt
„ Lucas Lawrence *	1st Co. 2nd Batt.	
„ R. B. Wilson	2nd Co. 2nd Batt.	(?)

* It is not certain whether these officers were with any portion of the force employed on service. They were absent from their companies and "in the field," but this expression did not necessarily imply active service.

NOTE C (*continued*).

Rank and Names.	Posting.	Remarks.
Lieut. J. Johnson	4th Co. 1st Batt.	
„ T. A. Vanrenen	3rd Co. 2nd Batt.	Detached with the Right Division from Agra in December, 1817
„ R. S. B. Morland	2nd Troop H.A.	Removed to 3rd Troop 15th December, but did duty with 5th Troop. Subsequently reverted to 2nd Troop
„ W. Geddes	Rocket Troop	
„ T. D'Oyly	7th Co. 2nd Batt.	
„ J. S. Kirby	3rd Co. 1st Batt.	Doing duty with 7th Co. 3rd Batt. Afterwards transferred to 6th Co. 2nd Batt., and in June, 1818, to 2nd Co. 2nd Batt.
„ T. Sanders	4th Co. 2nd Batt.	
„ R. R. Kempe	6th Co. 2nd Batt.	Afterwards 5th Troop H.A.
„ G. Twemlow *	6th Troop H.A.	
„ C. G. Dixon *	5th Co. 3rd Batt.	Quarter-Master Right Division
„ H. P. Hughes *	2nd Co. 3rd Batt.	Doing duty with 6th Co. 3rd Batt.
„ W. E. J. Counsell *	6th Co. 2nd Batt.	
„ J. H. Middleton *	1st Co. 1st Batt.	(?)
„ J. D. Crommelin * †	4th Co. 2nd Batt.	
„ C. R. Whinfield *	5th Co. 1st Batt.	
„ G. Pennington, jun.*	1st Troop H.A.	
„ Giles Emly * †	6th Co. 1st Batt.	(?) Afterwards 3rd Co. 2nd Batt.
„ A. Thompson *	1st Troop	Afterwards 2nd Troop
„ J. G. Barnard * †	1st Co. 3rd Batt.	(?)
„ T. B. Bingley * †	3rd Co. 1st Batt.	(?)
„ R. Burrowes *	2nd Co. 1st Batt.	Doing duty with 6th Co. 3rd Batt.
„ C. Patch *	7th Co. 1st Batt.	Doing duty with 6th Co. 3rd Batt. Transferred in June, 1818, to 5th Co. 2nd Batt. Died at Baitul 2nd Nov., 1818
„ T. Montgomerie *	3rd Troop H.A.	

* These officers were promoted to the rank here given them by the augmentation dated 1st September, 1816. The postings in all cases are those at the opening of the war.

† It is not certain whether these officers were with any portion of the force employed on service. They were absent from their companies and "in the field," but this expression did not necessarily imply active service.

NOTE D.

Officers of the Madras Artillery who served during the Pindári and Máhrátá war. N.B.—This list is not complete.

Lieutenant-Colonel ...	J. Crosdill, C.B.	
" ...	S. Dalrymple.	
Captain (Brevet-Major)	G. J. Goreham.	Died 20th May, 1818, at the siege of Chándá.
Major	J. Noble, C.B.	
"	A. Weldon.	
Captain	G. M. Poignand.	
"	W. Morrison.	
"	J. H. Frith.	
"	H. J. Rudyerd.	
"	B. Macintosh.	
"	S. Cleaveland.	
"	J. Maxwell.	
"	P. Poggenpohl.	
Lieutenant	J. G. Bonner.	
"	J. Bennett.	
"	N. Hunter.	
"	J. J. Gamage.	
"	W. F. Lewis.	
"	P. Montgomerie.	
"	G. Conran.	
"	A. F. Coull.	Wounded severely Dec., 1817, at the siege of Nágpur, and died 5th Nov., 1818, at Masulipatam.
"	J. M. Ley.	
"	T. G. Noble.	
"	J. T. Kelly.	Died 29th June, 1818, in camp Jálnah.
"	W. Chisholm.	Killed 1st Jan., 1818, at Korygám.
"	E. King.	Died 12th Nov., 1818.
"	F. F. Whinyates.	
"	F. Blundell.	
"	T. Cussans.	
"	Æ. Shirreff.	

Note E.

Officers of the Bombay Artillery who served during the Pindári and Máhrátá war of 1817, 1818, and 1819. N.B.—This list is not complete.

Lieutenant-Colonel		H. Hessman.
„		C. J. Bond.
Major ...	...	J. H. Pierce.
„ ...	...	G. B. Bellasis.
Captain	...	S. R. Strover.
„	...	E. Hardy.
„	...	R. Thew.
„	...	J. G. Griffith.
Lieutenant	...	T. Stevenson.
„	...	H. L. Osborne.
„	...	G. R. Lyons.
„	...	W. Jacob.
„	...	W. Miller.
„	...	J. Laurie.

Note F.

Officers of the Engineer Department who served during the Pindári and Máhrátá war.

Royal Engineers.

Lieutenant	...	J. H. Elliot.	Aide-de-camp to Lieutenant-General Sir T. Hislop. Served as an engineer officer at the siege of Talner.

Bengal Engineers.

Major ...	...	T. Anbury.	
Captain	...	Richard Tickell.	
„	...	J. Peckett.	
Lieutenant	...	J. Cheape.	
„	...	J. Colvin.	
„	...	A. Irvine.	
„	...	E. Garstin.	
„	...	J. F. Paton.	Adjutant.
Ensign	...	J. Warlow.	

MADRAS ENGINEERS.

Lieutenant	...	T. Davies.	Killed May 10th, 1818, at the siege of Máligám.
"	...	J. Coventry.	Died Dec. 8, 1821, of fever contracted during the Máhrátá war.
"	...	A. Anderson.	
"	...	A. Grant.	
"	...	— Macleod.	
"	...	J. W. Nattes.	Killed May 29th, 1818, at the siege of Máligám.
Ensign	...	J. Purton.	
"	...	J. Oliphant.	
"	...	— Underwood.	
"	...	E. Lake.	

BOMBAY ENGINEERS.

Captain	...	Justinian Nutt.
Lieutenant	...	T. Remon.

OFFICERS OFFICIATING AS ENGINEERS.

Lieutenant	...	Gavin R. Crawfurd (Bengal Artillery). At the siege of Chándá.
"	...	Ainsworth (34th Regiment). At the siege of Sholapore.
"	...	Wahab (Rifle Corps). " "

Note G.

Route from Sitabaldi, near Nágpur, to Siuni, showing the distance marched by the 3rd Native Troop, Bengal Horse Artillery, in pursuit of the Peshwa, April 16th to 17th, 1818, as verified in the office of the Surveyor-General of India.

Places.	Miles.	Remarks.
From Sitabaldi to Gumgáon	10½	Commenced march 1 a.m. April 16th
Tákalghát ...	8½	
Asola	4	
Kailjhur ...	7¾	
Ajangáon ...	3½	
Junáno	2¼	
Dailulgáon ...	4⅓	
Pápalgáon ...	½	
Dindora	2¾	
Madni	⅓	Cross Dhámna river
Alipur	11	
Alamdoh ...	2¼	Reached Col. Adam's camp between 2 and 3 p.m. Marched again at 9 p.m.
Kángáon ...	4¾	
Khángáon ...	4½	
Poti	2	
Bálágáo	3	Cross river Wardah
Pipalkuti ...	3¾	Must have been reached between 2 and 3 a.m. April 17th. Halted a short time, and pushed on with cavalry and horse artillery
Barora	2	
Boráti	4¾	Met with enemy near this place shortly after sunrise (5 a.m.)
Jatingdárá ...	3¾	
Moda	1½	
Asoli	3¾	
Siuni	3½	Encamped after the action, about 1 p.m. April 17th, at or near this place
Total miles ...	94$\frac{11}{12}$	

NOTE H.

Statement of ordnance employed in the attack upon the fort of Asirgárh, taken from Blacker's and other accounts.

	GUNS.						HOWITZERS.				MORTARS.		
	24-pr.	18-pr.	12-pr. iron.	12-pr. brass.	6-pr.	6-pr. gallopers.	8-inch.	5½-inch.	5½-inch on beds.	4⅖-inch.	10-inch.	8-inch.	5½-inch.
General Doveton's Division before the siege	...	2	...	...	...	...	...	2	—	—	—	—	—
One and a half troops horse artillery	...	...	...	3	...	6	—	—	—	—	—	—	—
General Malcolm and Captain Frith's camel battery	...	...	...	...	...	...	...	...	2	4	—	—	—
Jálnah battering train, joined 1st March	...	7	2	...	...	...	2	3	...	...	1	3	1
Nágpore „ „ 11th „	...	4	...	...	...	...	2	...	...	...	...	2	—
Hoshangábád „ „ 17th „	...	2	2	...	...	...	...	2	—	—	—	—	—
Ságar „ „ 3rd April	2	4	...	...	...	...	2	...	...	...	3	3	8
Total	2	19	4	3	...	6	6	7	2	4	4	8	9
As stated in Lake's "Sieges"	2	19	3	4	*	*	6	7	2	4	4	8	3
As stated in Captain Buckle's memoir	2	22 †	4	3	16 ‡	14	6	7	...	4	4	8	9

* Omitted, not being siege ordnance; Buckle's number probably is correct.
† A manifest error. N.B.—Prinsep gives the same numbers as Lake; the 5½-inch mortars of the latter are evidently wrong.
‡ Probably battalion guns.

Note I.

Return of casualties in the Bengal, Madras, and Bombay Artillery during the siege of Asirgárh.

	Killed.			Wounded.								
	Sub-Conductor.	Drummers.	Rank and File.	Major.	Captain.	Lieutenants.	Sergeants.	Rank and File.	1st Tindal.	Gun Lascars.	Sirdar.	Dooly-bearers.
Bengal Artillery ...	...	...	...	...	...	1	...	5	2	6	1	—
Hyderabad Subsidiary Force	...	1	1	1	1	...	1	8	...	8	...	1
Sir John Malcolm's Division, horse and foot artillery	1	1	1	...	...	...	1	7	...	7	—	—
Bombay Artillery ...	...	...	1	...	...	1	1	6	...	4	—	—
Total	1	2	3	1	1	2	3	26	2	25	1	1

Names of Officers Wounded: Major A. Weldon, Madras; Captain J. H. Frith, Madras; Lieutenant W. Counsell, Bengal; Lieutenant W. F. Lewis, Madras Artillery. All slightly.

Note K.

Description of the great Burhánpur gun, taken in the fort of Asirgárh (from the *Central India Provinces Gazetteer*).

Dimensions.

	ft.	in.
Length, muzzle to breach	12	9
" muzzle to trunnions	7	3
Circumference at breach	8	2¼
" in front of trunnions	6	6
" at muzzle	5	7
Diameter of bore	0	8¼

Inscriptions, beginning from the muzzle.

"When the sparks of sorrow issue from me, life deserts the body, as grief falls on the world when flames issue from the fiery zone."

"Abul Muzaffar Mohi ud din Muhammad Aurangzeb, Sháh Ghází." *

"Made at Burhánpur in the year 1074 A.H." (A.D. 1663).

"The gun Mulk Haibats." †

"In the rule of Muhammed Husain Arab."

"A ball of 35 seers and 12 seers of powder, Shah Jaháni weight."

---

* Emperor of Delhi. † Terror of the country.

NOTE L.

Officers of the Medical Department who served with the Bengal Artillery during the Pindári and Máhrátá war.

CENTRE DIVISION.

Surgeon James Macdowall.	Principal medical officer, and in charge of horse artillery
„ Colin Campbell.	In charge of foot artillery. Afterwards with Ságar Field Force.
Assistant-Surgeon Geo. G. Spilsbury.	With foot artillery.
„ R. B. Pennington.	With horse artillery.

LEFT DIVISION.

Assistant-Surgeon Hugh Smith, M.D.	Afterwards with Ságar Field Force.

RESERVE.

Assistant-Surgeon Charles Renny.	In charge.

NÁGPUR SUBSIDIARY FORCE (5TH DIVISION).

Assistant-Surgeon William Hastie.	In charge. Killed by the accidental discharge of his rifle in camp before Chándá.

CHAPTER XIII.

FIRST BURMESE WAR, 1824—1826—Hostilities previous to war being proclaimed—Captain Timbrell's flotilla—Operations in Assam—Lieutenants Bedingfield and Burlton—Operations in Káchár—Failure of attack on Dudpátli—Lieutenants Huthwaite and Smith join—Attack on Tiláyan—Brigadier-General Shuldham's force—Artillery detail—Move upon Manipur abandoned —Disaster at Rámu—Invasion of ÁRÁKÁN—General Morison's force—Artillery with it—Advances—Four columns formed—General McBean joins—Batteries opened—Brigadier Richards storms the Árákán heights—Great mortality in the force—It is broken up—RANGOON Expedition—Detail of force employed—Artillery—Capture of Rangoon—Want of supplies—Position taken up before Rangoon—Remarks thereon—First attack on Kemendine—Attack on stockades north of Rangoon—Second attack on Kemendine—British position attacked—Ten stockades captured—Tenasserim provinces reduced—Dalla—Tantabeng—Kaiklo—Martaban taken—Enemy concentrate—British position invested—Severe fighting for seven days—Total repulse of the enemy—Services of artillery officers noticed—Kok-keing—Reinforcements—Rocket troop—Lieutenant-Colonel G. Pollock—His energy—1st Troop, 1st Brigade H.A.—Second expedition to Tantabeng—Advance from Rangoon—Sir Archibald Campbell's column—General Cotton's column—Detatchment to Bassein—Sir A. Campbell countermarches on Donabyo—General Cotton attacks Panlang—But fails at Donabyo—Sir A. Campbell joins—Batteries open fire—Donabyo evacuated—Advance on Prome—Changes among artillery—Enemy invest Prome—Tsenbike — Napádi—Captain Lumsden wounded—Honourable mention of the artillery—Advance — Negotiations — Some artillery officers leave—Movements in Pegu—Sitang taken—Captain Dickenson hoñourably mentioned—Melloon—Its storm and capture—Honourable mention of the artillery—Advance—Action of Pagahm-Myo—Sir A. Campbell's order on the occasion—Peace.

WAR was proclaimed with the Burmese Government on the 5th of March, 1824. Previously to this, hostilities 1824 had already commenced on our north-eastern frontier. As early as September, 1822, a flotilla of gunboats was sent up the river Brahmaputra, to protect the line of frontier near Gwálpárá from violation. Captain Thomas Timbrell, who had lately returned from leave home, was appointed to command, with Lieutenants R. G. Bedingfield and P. B. Burlton as his subalterns. Captain Timbrell returned to Dumdum early next year, but the two latter officers remained in Assam.*

1824 Early in 1824 there was a small force stationed at Gwálpárá, under command of Brigadier George McMorine, which made some successful movements against the stockaded positions of the enemy at Gaoháti, Kaliábar, Hátbar, and Maura Mukh. Brigadier McMorine died of cholera early in May, and was succeeded by Lieutenant-Colonel Alfred Richards, the next senior. The setting in of the rains obliged this officer to fall back upon Gaoháti; but in the end of October he resumed operations. He had advanced to within a few miles of Rungpore, the capital of Upper Assam, when he was joined at Gaori Ságar, on the 25th of January, by Lieutenants Bedingfield and Burlton, with two howitzers and two 12-pounder carronades. The place was invested on the 29th, and the enemy driven within the fort, with some loss on both sides. On the 31st the place surrendered. Both Lieutenants Bedingfield and Burlton are mentioned with thanks in Lieutenant-Colonel Richards' despatch. Assam thus passed into our hands, and further hostilities were suspended on the line of the Brahmaputra.

When war became inevitable, the opinion of the commander-in-chief, Sir Edward Paget, had been taken

* See Note A in the appendix to this chapter, relative to these two officers.

as to the plan of operations; and he, pointing out the maritime borders of Burmah as fittest for offensive operations, warned the Government of the inadvisability of any other than a defensive policy on the eastern frontier. "Any military attempt beyond this," he wrote, "upon the internal dominions of the King of Ava he is inclined to deprecate, as instead of armies, fortresses, and cities, he is led to believe we should find nothing but jungle, pestilence, and famine"—a truth soon too truly realized.

1824

On the Silhet frontier, as well as in Assam, hostilities preceded war; the Burmese having, in the beginning of January, invaded Kachár, a province which Lord Amherst, the Governor-General, had found it necessary to take under British protection. It was assailed upon three sides—from the Kásiya hills on the west; from Assam, the district lying along the river Brahmaputra, on the north; and from Manipur on the east. At Silhet, the frontier station, Major T. Newton was at this time in command. He occupied advanced posts in Kachár, at Bhadrapur, Játrapur, and Tiláyan.

January

The force from Assam having, early in January, advanced as far as Bikrampur, Major Newton concentrated his force at Játrapur, and on the 17th attacked and carried the position. As he, however, immediately after withdrew his detachment to Silhet, the Burmese united their forces and recovered their lost ground, and occupied posts which they stockaded, even to within a thousand yards of Bhadrapur, where Captain Joseph Johnstone, with a small detachment of native infantry, was. This officer attacked and drove them out of their unfinished works, compelling the different divisions to fall back—that from the north upon the Bhartika pass, and the Manipur division upon Dudpátli.

February

1824 February

Lieutenant-Colonel Herbert Bowen, who had arrived and taken command, followed up this success by attacking and driving back into the hills the Assam division. He was not so fortunate, however, in his attempt upon Dudpátli. There was a strong stockaded work on the north bank of the Surma river, the rear resting on steep hills. Each face was defended by a deep ditch, fourteen feet wide, with a fence of spiked bamboos running along the outer edge; and on the land side, jungle and high grass covered the approach. On the 21st of February, after the battalion guns had opened with apparent effect, Lieutenant-Colonel Bowen ordered the assault on the western face; but the sepoys failed to make their way over the fences after several attacks repeated throughout the day. One officer (Lieutenant A. B. Armstrong, 14th N.I.) and 20 rank and file were killed; 3 European, 5 native officers, and 126 rank and file wounded. On the 27th of February, Brigadier W. Innes, C.B., with the 38th Native Infantry, joined at Játrapur, and assumed command. Lieutenant Edward Huthwaite, with a detachment of golandáz and four 6-pounder guns, being sent on by water, had arrived a day or two previously; and shortly afterwards Lieutenant Charles Smith, who had been ordered up from Calcutta, arrived with two 5½-inch howitzers, and took command of the artillery. The enemy retreated, and fell back upon Manipur, and for some time Brigadier Innes merely maintained a defensive attitude; but when, upon intelligence of the defeat of Captain Noton's detachment at

May

Rámu, he moved southward to cover Dacca and Chittagong, they advanced again and reoccupied the positions of Tiláyan, Dudpátli, and Játrapur. He therefore

June

returned to Silhet on the 12th of June, and shortly after moved into Kachár. The inundated state of the country made marching very difficult, and it was not till the 27th that Játrapur was reached. The hill of Tiláyan was occupied in great force and strongly stockaded. From this the brigadier endeavoured to dislodge the enemy, but failing to do so, he desisted on the 8th of July, and fell back, his force not being sufficient to occupy all the heights which commanded the point of attack. In his despatch, dated July 9, to the adjutant-general, he says:—

1824 June

July

"I have however to observe, that this measure was determined on before, in consequence of the exhausted state both of the artillery and infantry of my detachment, the former having been in the batteries from the morning of the 6th until the 8th instant. I think it my duty to bring to the notice of his Excellency the very zealous exertions of this arm of the service; the practice was beyond praise, and the shot and shells were thrown with a precision that could not be surpassed, but the 6-pounder shot were found to have no effect on the enemy's works, although the shells must have done considerable execution.

"I feel myself much indebted to Captain Smith for his great exertions during the three days the battery was open, and to Lieutenant Huthwaite, who, though labouring under a severe fever, rendered me the most essential service."

Fever had already begun to put a stop to all active operations. Lieutenant Huthwaite was sent to sea for the recovery of his health, but Captain Smith remained in command of the artillery till December. No further movements took place on either side during the rains.

December

Subsequently, however, a force was assembled and placed under command of Brigadier-General Thomas Shuldham, consisting of two brigades, numbered respectively the 3rd and 4th, to which were attached four companies of pioneers under Major Swinton, and detachments of the 6th or Golandáz Battalion, commanded by Captains Jonathan Scott and Charles Smith, with Lieutenants J. Turton (adjutant), F. Brind, and J. T. Lane, subalterns.

1825

1825 This force was intended to march upon Manipur, with the view of, at least, diverting the attention of the Burmese Government from the principal points of attack. But between Silhet and that place, a distance of 140 miles, intervened several parallel mountain ridges running north and south, and covered for many miles with dense forest and soft alluvial soil. Heavy and constant rain made the ground quagmire, and the February rivulets torrents. During the whole of February, 1825, the pioneers were constructing a road through the forest to a nullah about forty miles west of Bánskandi; but their labours were all thrown away. No road passable for guns could be made; elephants, camels, and bullocks perished in conveying supplies to the working parties; and on the 23rd of March General Shuldham reported to Lieutenant-Colonel Nicol, the adjutant-general of the Bengal army, the impossibility of keeping any force eastward of the forest which was so formidable a barrier, and the force was in consequence broken up.

1824 At Chittagong, further south, Colonel John Shapland, C.B., commanded a force of about 2000 native infantry when the war broke out. A large Burmese force was assembled in Árákán, under the command of Máhá Mengi Bandula, an officer of great reputation at the court of Ava, and who had been one of the foremost counsellors for war. To watch any demonstration, Captain Thomas Noton, 45th N.I., was stationed at Rámu with five companies of his regiment, some newly raised local irregulars, and two guns under Lieutenant James Warner Scott. This outpost was distant 70 miles by land from its main body, and was without cavalry or any intermediate support; hence the disaster which May occurred. On the 10th of May, news was brought that the Burmese were advancing upon Ratnapalling, an outlying picquet distant about ten miles, and Captain Noton went forth to reconnoitre with all his disposable force; but the move was injudicious and unfortunate. The new levies would not fight; one of the elephants got restive and threw the 6-pounder he carried; another got rid of his load of ammunition, and the mahouts could not or would not control them; the detail of golandáz was too small to perform the ordinary service of loading and unloading the animals, and the ordnance could not be brought into action.

1824 May

So Captain Noton had to return without effecting anything. On the 13th the enemy had taken up a position close to Rámu, and commenced entrenching themselves, a precaution which Captain Noton does not appear to have taken on his part. Had his post been properly secured, there is some reason for thinking that he might have been able to hold his ground till reinforced, as he hoped to have been. On the 15th the enemy crossed a river which separated them from the position, and commenced an attack, which was kept up by night as well as by day. A tank in the rear, surrounded by a high embankment commanding the main post, was held by a strong detachment of the irregulars; but these, on the morning of the 17th, lost heart, and were driven from it. Captain Noton was obliged to abandon his guns and retreat. His men, who, except the irregular levies, till now had behaved very well, pressed hard, and wearied with the previous fatigue and privation, soon dispersed, and all the European officers except three were killed. Lieutenant J. W. Scott, who had been severely wounded, had been tied, by Captain Noton's orders, on an elephant, which, with the rest, fled in alarm; and he was thus saved. Lieutenant Robert Coddrington and Ensign Campbell, N.I., were the two others.

1824 May

Fortunately, however, the enemy were contented with their success. Colonel Shapland's force was immediately strengthened, and this part of the frontier secured. Máhá Bandula was soon afterwards recalled for the protection of the Ava provinces on the Rangoon side.

September

To co-operate with the expedition which had already landed, and had occupied but not advanced beyond Rangoon, a force was assembled in September at Chittagong, under command of Brigadier-General J. W. Morrison, C.B., H.M.'s service.

The artillery with this portion of the army was composed as follows:—

Lieutenant-Colonel Alexander Lindsay (Bengal), commanding.
Lieutenant John S. Kirby (Bengal), Adjutant.

4-22 R.A. 3rd Company, 2nd Battalion * (Bengal).—Captain John Rawlins; Lieutenants James R. Greene and George Hart Dyke; 2nd Lieutenants W. C. J. Lewin, Henry M. Laurence, and Samuel W. Fenning.

C-19 R.A. 4th Company, 2nd Battalion * (Bengal).—Captain Edward Hall; Lieutenant John Hotham; 2nd Lieutenants John Fordyce and Ambrose Cardew.

1-17 R.A. A Company, 2nd Battalion † (Madras).—Captain John Lamb; Lieutenant George Middlecoat.

Besides the land force, which amounted altogether to about 11,000 men, there was a flotilla of gunboats under Commodore Hayes, of the Indian navy. The first object of this army was Árákán, the capital of the district bearing the same name, and from thence to cross the range of mountains to the east, and descend into the Irawádi valley at Sem-byo-gyun. To reach Árákán, General Morrison preferred the coast to the overland route, because the latter, with its mountains and roadless

* These companies were then designated the 6th and 7th Companies, but the numbers were altered by G. G. O. 24th June, 1825, to those given in the text.

† The number of this company was A Company, 1st Battalion; it was changed by a regimental order dated 1st September, 1825, to that given in the text. It reached the coast of Árákán in June.

jungle, appeared to be impracticable. But the flotilla could surely have transported without difficulty, and in much less time, the whole of the ordnance and most of the other stores, with part of the troops, to the north of the Árákán river. It was not till January, 1825, that the general, delayed by want of carriage * and late rains, moved from Chittagong; and on the 1st of February he reached Tek Náf. From this place progress was very slow; transporting over the wide estuaries of the rivers was very laborious. On the 22nd of February, Commodore Hayes, with a detachment of the flotilla and some troops on board, stood up the western branch of the Árákán river, and made an unsuccessful attack upon the stockades at Chambálla. It was not until the 24th of March, however, that the army began to move up the eastern bank of the main branch. On the 26th, four columns were formed as follows:—

1824

1825 January

February

March

RIGHT COLUMN.

Brigadier Grant, C.B.

54th Regiment; 42nd and 62nd N.I.; one company N.L.I., and one of pioneers.

Ordnance.—One 12-pounder and two 6-pounder guns.

* It was this want of carriage that was the main cause of "the Barrackpore mutiny," not very erroneously described as an ebullition of despair on the part of the sepoys at being ordered to march without the means of doing so. The 26th, 47th, and 62nd N.I., stationed at Barrackpore, received in October orders to march for Árákán. Carriage was not to be had, and Government refused to supply it. The disaster at Rámu, a dread of the Burmese, a suspicion that they might be sent to Árákán by sea, intensified the just cause they had for complaint. The 47th being the first ordered to march, broke out into open mutiny on the 1st of November, 1824. The 1st (Royal Scots) and 47th Regiments and a troop of the body-guard were marched to Barrackpore. During the night Captain N. S. Webb's Horse Field Battery was brought over from Dumdum. His two subalterns were Lieutenants Macvitie and Pillans. At daybreak on the 2nd, the mutinous regiment was paraded before the commander-in-chief, in front of the two European corps, the battery somewhat in rear. The sepoys were ordered to march immediately, or ground their arms. They stood sullen and unyielding till the guns opened fire upon them, when they broke and fled.

1825 March

CENTRE COLUMN.

Brigadier Richards.

44th Regiment; 26th and 49th N.I.; two companies N.L.I., and one of pioneers.

Ordnance.—Two 12-pounder guns and two 5½-inch howitzers, Lieutenant-Colonel Lindsay commanding.

LEFT COLUMN.

Captain Leslie, 54th Regiment.

(To act with the gunboats).—Two companies 54th Regiment; two N.L.I.; one of the Magh levy, and two of Magh Pioneers.

RESERVE COLUMN.

Lieutenant-Colonel Walker.

Three companies of the 54th Regiment, five of native infantry, and four of pioneers, with the 2nd Local Horse.

Ordnance.—One 12-pounder and two 6-pounder guns.

The left column proceeded up the main channel, but the boats grounding, it was landed, and began to skirt the river with the view of turning the enemy's right. The right and centre columns, ascending the hills, dislodged and drove them back to the works that covered the fords at Maháti. This occupied nine hours, and the troops bivouacked within a mile and a half of the enemy's principal post. The reserve column, detained in bringing on the principal part of the artillery, joined at midnight.

On the 27th the force advanced against and took the post of Maháti; the guns (four 12-pounders, two 6-pounders, and two 5½-inch howitzers) opened fire, but the enemy did not wait long to judge of their effect.

Next day the force halted to allow the rear to close up. Brigadier-General McBean, who had come from Rangoon, joined to-day with the 5th Brigade, the 11th and 16th Madras Native Infantry, commanded by Brigadier Fair, of the same service.

From this place one pass only led to Árákán, and this was difficult and strongly fortified. Brigadier-General McBean, with the 5th Brigade, six companies of the 54th, and some native infantry, attacked the pass in front; but

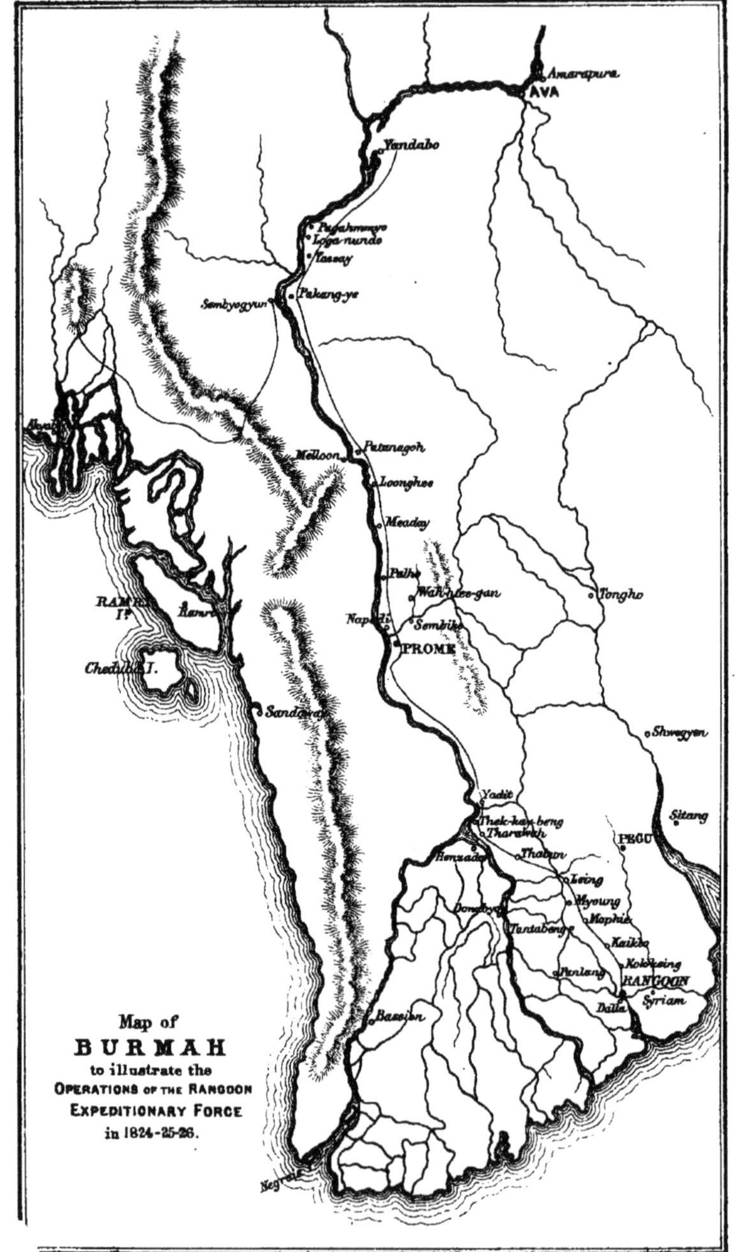

London Henry S. King & Co. 65, Cornhill

Edwd Weller, Litho.

the steepness of the ascent and the constant shower of stones forbade the attempt, which was persevered in till every officer was wounded. Captain Lamb, having advanced his guns too far to cover the movement, was at last obliged to leave them behind.

1825 March

The next day, the 30th, was occupied in establishing batteries for four mortars, two 24-pounders, four 12-pounders, and two 5½-inch howitzers, which opened on the morning of the 31st, and kept up during the day a heavy fire. During the night Brigadier Richards was detached, with two columns, to attack the enemy's right upon the way for the enemy to Árákán. As he had done ten years before at Jaitak, he performed this duty well; the heights were gained ere the enemy detected the move, and, panic-struck, they gave way before the combined attack. Árákán was taken. The casualties among the artillery in these operations were—Killed: 1 bombardier; Wounded: 5 rank and file, and 5 drivers.

April

But, except the capture of Rámri * and Sandoway by detachments, this was all that General Morrison's force was destined to perform. Several reconnaissances were made, but they failed to find a practicable route across the hills, by which the force could unite with Sir Archibald Campbell. Meanwhile, fever and dysentery broke out with so much violence that Government was compelled to break up the force, already deprived of all vitality as an army.

April to September

Rangoon Expedition.

In March, 1824, the Government of India had determined upon sending an expedition from the Presidencies of Fort William and Fort St. George to attack Burmah

1824

* Captain Hall commanded a small detail of artillery at the capture of Rámri, in February, 1825. He was obliged to return to Cheduba for his health shortly afterwards.

1824 from the side of the province of Rangoon. The command of the united forces, amounting to 12,845 men,* was given to Brigadier-General Sir Archibald Campbell, K.C.B., of the 38th Foot, who had served in the Peninsular War with distinction. The artillery, drawn from the three presidencies, was constituted as follows:—

Lieutenant-Colonel C. Hopkinson (Madras), commanding (joined in October.)†

Major W. M. Burton (Madras).

Captain P. Montgomerie (Madras), Brigade-Major.

Lieutenant W. F. Lewis (Madras), Commissary of Stores.

Lieutenant-Colonel G. Pollock, commanding Bengal Artillery (did not join until December).

Lieutenant G. H. Rawlinson, Adjutant to the Bengal detachment. Officiating in the Commissariat.

Lieutenant G. S. Lawrenson, Officiating Adjutant to the Bengal detachment.

2-25 R.A. Reduced in 1871 — 3rd Company, 5th Battalion (Bengal).—Captain Thomas Timbrell; Lieutenants George R. Scott and R. G. Macgregor.

D-19 R.A. — 4th Company, 5th Battalion (Bengal).—Captain Edward Biddulph (did not join until December); Lieutenants W. Counsell and Errol Blake; 2nd Lieutenants E. F. O'Hanlon and James H. Macdonald.

B-20 R.A. — B Company, 2nd Battalion (Madras).

A Company, 5th Battalion (Madras) Golandáz.‡ The names of the officers are given in the appendix to this chapter, Note E.

A detail of Bombay Artillery, probably one company.—Captain Lechmere C. Russell.

---

* See detail in the appendix to this chapter, Note F.

† Major W. M. Burton, at first commanding, was superseded by Lieutenant-Colonel Hopkinson. G.O.C.C. Fort St. George, Oct. 7th, 1824.

‡ The conduct of the golandáz in preparing for service beyond sea contrasted strongly with the spirit shown by the Bengal sepoys, to which allusion has been made. It is true that they had not the same cause of complaint with respect to carriage. The golandáz for Madras were recruited at this time from the neighbourhood of Benares. They were asked to volunteer, and did so to a man, and they made no difficulty with regard to their provisions on board ship. This is on the authority of Colonel R. S. Seton, who had ample opportunities of knowing the spirit of the men there. The Madras sepoys, too, evinced unusual ardour, as Sir Thomas Munro testified (Minute dated Fort St. George, 18th June, 1824). Away from his own presidency, the Bengal Hindu was not infected by bad example.

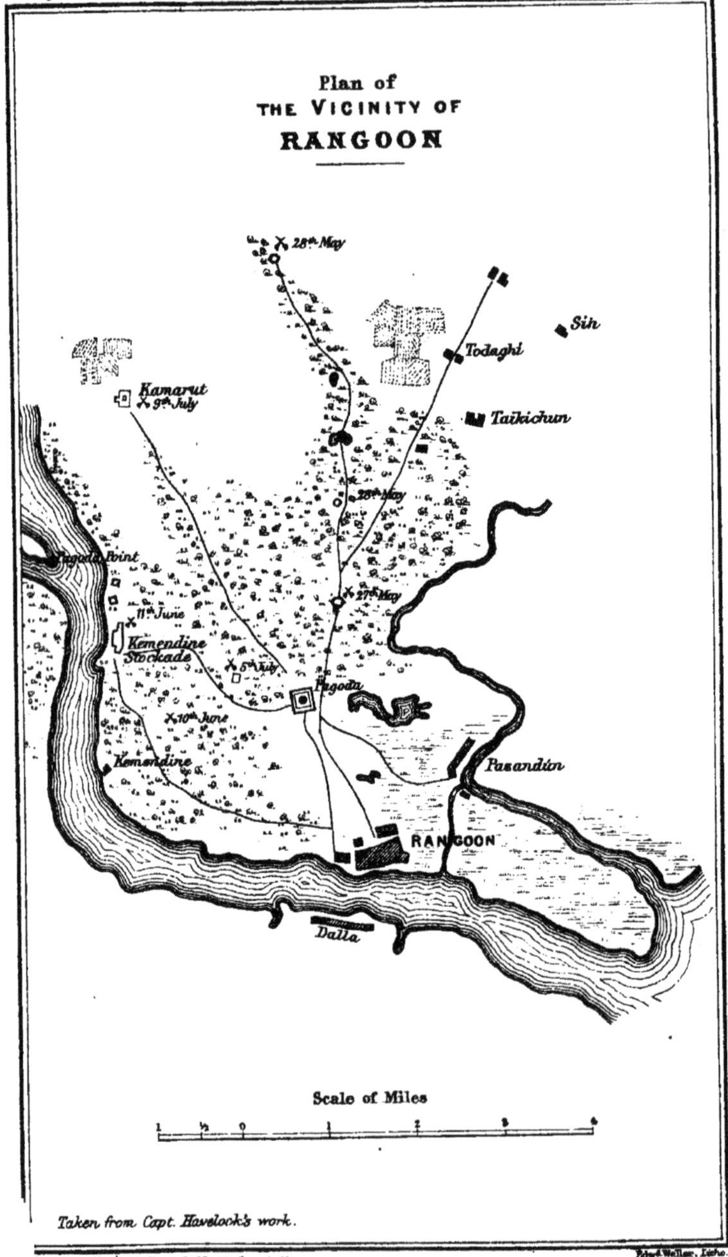

Taken from Capt. Havelock's work.

London: Henry S. King & Co. 65, Cornhill

1824 May

The fleets from Madras and Calcutta met at Port Cornwallis, in the Andaman Islands, and had nearly all arrived in the first three days of May. Two expeditions against the Islands of Chedula and of Negrais having been despatched under Brigadier McCreagh, 15th Foot, and Major James Wahab, 12th Madras N.I., Sir A. Campbell sailed with the remainder for Rangoon. They arrived off the mouth of the Rangoon river on the 9th, anchored within the bar next day, and on the morning of the 11th proceeded up with the tide, H.M.'s frigate *Liffey*, Commodore Grant, and the sloop *Larne*, Captain Marryat, leading. But little opposition was offered, and the transports one by one took up their position before the town. A feeble fire was soon silenced by the guns of the *Liffey*; and when the troops landed, the Burmese, terror-stricken, fled before them, and before many minutes Rangoon was in the possession of our troops. But it was deserted, and instead of the extensive supplies expected from a place of considerable commercial activity, little or nothing was found. It was soon discovered that the Burmese authorities, surprised as they were, had swept all the neighbouring plains of their herds and the villages of food; and thus was laid the foundation for much of the privation and disease endured by the troops in the first part of the war. For the Bengal and Madras troops had only a few weeks' supply of provisions, and the salt meat and biscuit on which they had to depend for some time after, even had these been good, would not at any time have sufficed to keep them in health; so that the fever incidental to the rainy season became endemic, though its form was generally a mild one, and, with rheumatism and dysentery, reduced the army to such a state, that at one time scarce 300 men were fit for duty. In the artillery, from June till

1824 May

October, the average monthly admissions into hospital were 65 Europeans and 62 natives, nearly one-third of the greatest numerical strength of the former, and one-fourth of the latter; yet this was a smaller proportion than in any other European regiment in the force.[1]

The positions occupied by the British army will be seen from the accompanying sketch. Two roads running from the north and western gates converge towards the

See PLATE. No. XLIV.

VIEW OF THE SHWE-DA-GON PAGODA, RANGOON.
(*Sketched in* 1873 *by Major J. B. Richardson, R.A.*)

great Shwe-da-gon pagoda, distant about a mile and a half from the town. The troops were cantoned, facing outwards, in the various buildings, sacred and profane,

[1] Transactions of the Medical and Physical Society of Calcutta, vol. iii., quoted by Wilson, p. 52.

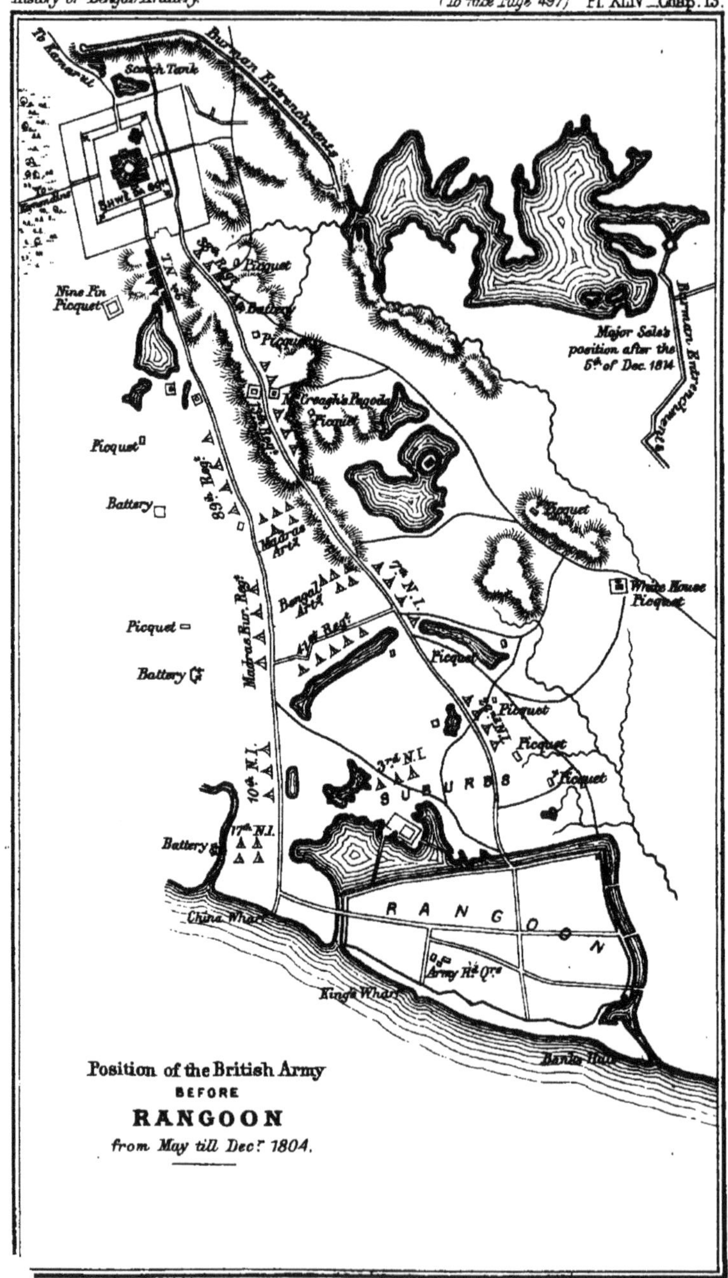
To Kamaroot
Scotch Tank
Burman Entrenchments
Shwe Da gon
Nine Pin Picquet
Picquet
Battery
Major Sale's position after the 5th of Dec. 1814
McCreagh's Pagoda
Picquet
Battery
89th Regt
Madras Artil.
Bengal Artil.
41st Regt
Madras Eur. Regt
Picquet
Battery
White House Picquet
Picquet
Picquet
Picquet
Picquet
Picquet
37th N.I.
10th N.I.
17th N.I.
Battery
SUBURBS
China Wharf
RANGOON
Army Hd. Qrs
Kings Wharf
Position of the British Army
BEFORE
RANGOON
from May till Decr 1804.

1824 May

along these roads. The great pagoda, in itself a fortress, was held by a battalion of Europeans. Standing on an abruptly rising eminence, in the centre of a platform revêted with brick masonry, about 75 feet higher than the road below, with an area of about two acres, the building itself rose 133 feet, and was surrounded by temples and the houses of the priests, in which our soldiers found a lodging. It was the key of the British position.

A writer * who has ably described the military events of this war, while he confesses that the British general was not wrong in thus taking up a position before Rangoon, yet questions whether an advance might not at once have been made before the enemy could have collected their forces, and thus, by a daring stroke of generalship, the command of resources of which he stood so much in need have been secured. But on reviewing the matter, taking into account the extreme deficiency of provisions, the total absence of means of transport,† except what the smaller craft of the flotilla afforded, and the impossibility of keeping his communication with the nearest seaboard open, as long as there were any enemies in his rear or upon his flanks, the English general must be held to have followed the only course open to him. The error of sending a force to operate in a country of forest and swamp, with such a rainfall as Burmah, just before the rainy and unhealthy season set in—while it, exclusive of other considerations, forced upon him defensive measures —was not his fault.

The form of the position taken up—two lines, each about two miles in length, forming an acute salient angle

* Havelock, p. 43.

† No cattle even for the guns had been sent, and a few found in Rangoon, of a very inferior breed, were all the artillery had to commence with.

1824 May

in front of Rangoon—was also determined by circumstances. For the dense forest which covered the country beyond and on either side of the great pagoda, especially to the west of it, prohibited the establishment of a line with its flanks resting on the Rangoon river and the Pazandun creek.

The enemy received reinforcements towards the latter part of May, and commenced drawing round the presumptuous invaders a line of works previous to driving them back into the sea. On the 16th a small force was sent up the Rangoon river, which landed and demolished a work commenced by the enemy at Kemendine. On the 27th a small party of the 38th Regiment, only 18 men, about a mile north-east of the pagoda, attacked and took a stockade, entering by an opening which had not been closed. The military secretary, Captain Snodgrass, was present. It was evident the enemy were in force in this direction, and not far off.

Next morning, therefore, Sir A. Campbell moved out with two companies of the 13th, two of the 38th Regiment, and 250 sepoys, a gun and a howitzer; Captain Timbrell and 2nd Lieutenant J. H. Macdonald with the latter. Rain had been falling since the 13th, and the country, forest and field, was under water. The country-bred cattle were unable to drag the guns through the swampy ground, so the artillerymen were soon obliged to take off shoes and stockings, tuck up their trousers as high as possible, and assist with the drag-ropes. In this way they came to the stockade captured the evening before, which had been again occupied, but made little resistance. A mile further, two more, evacuated as they advanced, were destroyed. Every now and then openings in the forest showed the retreating enemy, and afforded an opportunity for some excellent practice with shot and shrapnel. But at seven miles from camp, both men and cattle with the guns could do no more, and with Lieutenant Macdonald were sent back to camp, escorted by the native infantry, Captain Timbrell going on with the general. Notwithstanding the rain, the little force went on some two miles further, when they suddenly came upon two stockades at the edge of a piece of jungle, which opened upon them. Brigadier-General McBean, with one company of the 13th, kept the plain, while the others advanced to the assault.

1824 May

The 38th, led by Major Evans, took one stockade, and the company of the 13th, under Major Dennie, the other, though stoutly defended by very superior numbers. It was not the last time that these gallant regiments strove together for the prize of war. The laurels gained by the 13th afterwards in Afghánistán, were worthily engrafted upon the honours it won in Burmah. The enemy first learned on the 28th of May how British soldiers could win a stockade without the aid of artillery.

June

The next affair, however, was not so successful. It has been laconically described by Havelock in the following words:—

> "On the 3rd of June the British attacked Kemendine by land and water. They did not manœuvre skilfully. An unlucky incident operated to their disadvantage; they made many bold efforts, but were repulsed."

In fact, the place had been attacked by a naval force from the river, in combination with three converging squadrons on land. One column was unable to reach its destination; of the others, one came under the fire of the naval squadron, and both failed in securing an entrance. An attempt at negotiation was made by a Burmese envoy on the 9th, but it came to nothing; and on the 10th, the general repeated his attack on Kemendine; this time with success.

1824 June

Long before daylight, a force of 3000 men was in motion towards the place. The Bengal Artillery was detailed for this duty, most of its officers present being sent.* The ordnance taken was four 18-pounders, four mortars, two 6-pounder guns, and two howitzers. The field-pieces were drawn by bullocks; the heavy guns by details from the infantry. About two miles from Rangoon, they came upon a stockade. The general called up two of the heavy guns and the field-pieces. They opened within fifty yards, but the round shot made little impression on the bamboo defences, passing through without cutting them down. At the end of an hour, however, a gap had been made, and the general ordered the assault. Led by the Madras Pioneers, who commenced to demolish the *abatis* and fill up the *trous-de-loup*, a part of the Madras European Regiment advanced, supported by the 41st, under the fire which the enemy then commenced to pour in. Another column of the 13th and 38th went round to the rear, and thence assaulted the high palisades. They had no ladders; but Major Sale, the Sale of Jelalabad, got himself raised up on the shoulders of his men, and was first in. Attacked on two sides, the Burmese soldiers knew not where to turn, and wildly jumped into the midst of their enemies, or rushed back into the middle of the work, there to be bayoneted. In ten minutes afterwards the place was won, and some 150 of its garrison killed, with a loss on our side of two killed and forty-eight wounded.

The delay involved in this affair, and the heavy rain which fell and continued all night, made it late when the force arrived in front of Kemendine. It was

* It also appears from the Madras Artillery Records, and Begbie, that Major Burton was present; but this officer's name does not occur in any despatch; he was shortly after superseded by Lieutenant-Colonel Hopkinson, and returned to Madras.

1824 June

a very extensive position, and could not entirely be invested, but communication was kept up with a fleet of gunboats sent up the river during the night.

Notwithstanding the weather, the unsheltered troops constructed batteries, and as soon as the light permitted, a fire both from mortars and guns was opened. Not satified with the little effect produced by the round shot upon the bamboo palisades, the general ordered up Lieutenant Macdonald with two 18-pounders, covered by a detachment of the 13th Regiment, to within twenty yards of the place; and these with a few rounds of bar shot * made an opening. Orders were given for the advance of two columns, already prepared for assault. Some of the gunners rushing forward, albeit without leave, were the first to enter; but only to find the last of the garrison escaping by their boats from the river side of the stockade.† A strong position was thus won without loss; and although a brave soldier[1] may have regretted that an assault was not delivered the evening before, and severer punishment not inflicted upon the enemy, the voice of reason will commend the general who gained his point without losing his men, while his previous successes were still fresh in the minds of his foes. It is certain that the assault on that day would not have been by any means bloodless, had not the artillery fire, that of the mortars especially, driven all thoughts of resistance out of their heads.

Leaving a detachment of the Madras European

* It is hardly necessary to say that bar shot had long before ceased to form part of any gun equipment, but it was thought that they might prove useful against bamboo stockades. A small proportion therefore was sent to Burmah.

† The details of this affair are taken from the despatch, Havelock, Snodgrass, and a very circumstantial account in a letter from the late Lieutenant-Colonel J. H. Macdonald.

[1] Havelock, pp. 78, 79.

1824 June

Regiment, with some native infantry and a detail of Madras Artillery, to garrison this post, under command of Major C. W. Yates, 26th Madras N.I., the general returned to head-quarters. The force sent against Cheduba returned the same day, after having driven the Burmese out of that island, and having left Lieutenant-Colonel R. Hampton with the 40th N.I. to garrison it. A detail of Bengal Artillery with two 9-pounder guns, under a sergeant, was with this force. Major Wahab's force returned from Negrais a few days after. They had found a barren island, not worth the fighting they had had for it. A detail of Madras Artillery was with them.

The arrival of these two detachments and of the 89th Regiment from Madras was a welcome addition to the British force, now suffering very much from fever and rheumatism, brought on by the exposure and privations the men had undergone. The enemy, also reinforced, began again to blockade the invaders. On the 1st of July they made an attack to the right and front of the great pagoda, penetrating between the British position and the village of Pazandun, which they set on fire, and attacked the right of the Bengal line. They were checked by the fire of the ship carronades, which were in the advanced picquets, manned by artillerymen, and driven back by a bold charge of four companies of the 43rd Madras N.I., commanded by Captain Jones, supported by a gun and howitzer of the Bengal Artillery.

July

Foiled in this attempt, they commenced erecting a strong stockade about half a mile from the pagoda, in the direction of Kemendine. This was taken on the 5th. Further to the north-east they had erected at Kamarut, near the banks of the river, in about six days, strong stockades capable of holding about 12,000 men, besides works at Pagoda Point, at the junction of the Leing branch with the Rangoon river, as well as on either bank of the stream a little lower down.

1824 July

On the 8th a combined attack was made on both positions. Sir A. Campbell accompanied the river party against the works at Pagoda Point; Captain Marryat had made all the naval arrangements. Major James Wahab commanded a detail of Madras N.I.; and Lieutenant-Colonel Henry Goodwin, afterwards the leader in the second Burmese war, a detachment of the 41st Regiment and one company of the Madras European Regiment. Brigadier-General McBean commanded the land attack upon Kamarut: 250 from each of the 13th, 38th, 89th Regiments, Madras European Regiment, and 7th Madras N.I., Madras Artillery accompanying them. The incessant rain which had fallen made the country almost impassable, and the road, or rather pathway, through the jungle was so narrow that at last all the guns had to be sent back, only a few small howitzers, carried by coolies enlisted and sent from Madras, being taken on. The operations were crowned with the most brilliant success. The 13th and 38th, under Majors Sale and Frith, were appointed to lead the assault on Kamarut, and as usual distinguished themselves. Seven stockades fell in succession, making with the three taken on the river, ten in all. Thirty-eight pieces of ordnance, 40 swivels, 300 muskets, and upwards of 800 killed, were cheaply purchased with the loss on our side of four rank and file killed; one officer (Captain Johnson, 13th) and 35 rank and file wounded.

For some time after the Burmese were not inclined to try conclusions with our troops. On the 5th of August a small force was sent against Syriam, on the Pegu river, under Lieutenant-Colonel Smelt; and on the

August

1824 August

8th another,* under Lieutenant-Colonel Hastings M. Kelly, of the Madras European Regiment, to attack some stockades up the Dalla creek. Portions of the Bombay Artillery accompanied both columns. In the latter service, the troops had to wade through shallow water to the point of escalade; and the officers, forming line breast-deep in mud and water, passed the scaling ladders from one to another, to be planted against the stockade.[1]

On the 26th of August a small force, consisting of the 89th Regiment and 7th Madras N.I., the Bombay Artillery under Captain L. C. Russell, was sent against the coast of Tenasserim under command of Lieutenant-Colonel E. Miles, C.B., with the view of endeavouring to bring the court of Ava to reason by depriving it of this portion of its territory. Tavoy was taken on the 9th of September, and Mergui on the 6th of October, and the whole coast gladly placed itself under British protection.

Brigadier-General McBean left this division of the army for that forming under command of General Morrison in the month of August, and was succeeded by Colonel Hugh Fraser, of the Madras army.

September

Meanwhile, as the enemy were becoming again troublesome in the vicinity of Dalla, a force proceeded up that creek on the 2nd of September, under Major R. L. Evans, Madras N.I. Captains T. Y. B. Lennan† with some howitzers, and Timbrell with two mortars, accompanied it. The general reports that the excellent practice of the artillery and gunboats soon forced the

* Details from H.M.'s ship *Larne*: Madras European Regiment, 18th and 34th Regiments Madras N.I., and 1st Battalion Pioneers.

† This officer, though two years junior by length of service, was senior to Captain Timbrell in rank by six months.

[1] Sir A. Campbell's despatch, dated 11th August.

enemy to abandon their defences with loss. Our casualties were one man wounded.

1824 September

October

On the 5th of October Major Thomas Evans, 38th Regiment, was sent with 300 men of that corps, 100 of the 18th Madras N.I., and a detail of Bengal Artillery under Captain Timbrell, on board a squadron of gunboats commanded by Captain Chads, H.M.'s ship *Arachne*, to proceed up the Leing branch of the river. They reached on the 7th the village of Tantabeng, which was defended by a stockade, breast-works, and a fleet of war-boats. Landing, the troops attacked and took the stockade, where they discovered a large quantity of powder and petroleum,* all of which being destroyed and the ordnance taken away, they returned to Rangoon. Captain Timbrell's name appears among those thanked for their services on this occasion.

The next event was not a success. Lieutenant-Col. H. F. Smith, C.B., commanding the 4th Madras Brigade, was sent with the 3rd and 34th Madras N.I. to attack the positions the enemy had taken up at Todaghi and Kaiklo.

Captain A. L. Murray and Lieutenant Aldritt, of the Madras Artillery, with two 4⅖-inch howitzers carried on camels (subsequently increased to four), accompanied them. The first stockade at Todaghi fell on the 5th; after which, being reinforced by 300 native infantry with two howitzers, they proceeded on the 7th to Kaiklo. Here they failed, and were obliged to retire with much loss. Had they been supported even by a small body of Europeans, the services elsewhere performed by the same regiments show that they would have been more fortunate on this occasion. Sir A. Campbell notes with praise, in his despatch of the 11th October, the efforts of

* Used in the construction of the fire rafts which the Burmese occasionally floated down the river.

1824 October

the British officers to lead and rally their men; and Colonel Smith singles out for commendation in his letter of the 10th the names, among others, of the two artillery officers,* and mentions the steadiness of the gunners under a galling fire, from which they appear to have suffered severely.

The general lost no time in repairing this disaster. On the afternoon of the 9th, Brigadier McCreagh marched out with 420 Europeans, 350 native infantry, and a detail of the Bengal Artillery under Lieutenant G. S. Lawrenson, a 5½-inch mortar, a 5½-inch howitzer, and a 6-pounder gun. A sunstroke on the first day did not deter the brigadier from remaining, though in a dooly, at the head of his men. Todaghi was reached at seven o'clock on the morning of the 10th. After a halt, leaving a detachment to occupy the stockades here, he took up his ground at sunset within a mile of the position at Kaiklo. On the road they passed the bodies of officers and men who had fallen on the 7th, crucified or mutilated with savage and indecent barbarity. It was well, perhaps, that the Burmese did not await the encounter with our soldiers; but post after post was evacuated. They retired on the position at Kágháhai. Securing Kaiklo with some native infantry, the brigadier moved on with the rest of his force and the artillery on the morning of the 11th; and next day found the place deserted, the enemy still in retreat. After destroying as much of the defensive works as could be done without a further halt, the force returned to Rangoon. In his despatch of the 14th October, the brigadier says:—

* "Captain Murray and Lieutenant Aldritt, of the Madras Artillery, were from the first zealous and indefatigable in their exertions in bringing their howitzers to the positions fixed upon; and the steadiness and alacrity evinced by them and their men, under a galling fire, was such as has on all occasions distinguished that corps."

"The manner in which the Bengal Artillery was forced over the most unfavourable ground and various difficult obstacles reflects high credit on Lieutenant Lawrenson and his detachment."

1824 October

On the 13th of October Lieutenant-Colonel Goodwin left Rangoon, with the 41st Regiment and the 3rd Madras N.I., to attack Martaban. Captain Kennan commanded the artillery, and Lieutenant R. G. Macgregor, with part of his company, was also present. Calms and counter-currents delayed progress, and it was not until the 29th that the fleet arrived before the place. It is situated at the bottom of a high hill, and its defences, both of timber and masonry, looked formidable. During the night a cannonade was kept up on both sides; and the excellent practice made by Captain Kennan, and by Lieutenant Macgregor in the bomb-vessel, is mentioned in the despatch. Next morning the place was stormed and taken. Lieutenant Macgregor, with eight of his men armed with muskets, were of the attacking party, and were distinguished in the attack.[1] The casualties in the artillery were one gunner killed and two wounded (Madras). The subsequent operations of this force were not of any great importance as to detail, though it did not rejoin the army till the month of December.

November

After the cessation of the rains at the end of October, the health of the troops improved considerably, although few regiments could muster more than two hundred and fifty sickly men under arms. Still the spirit that animated them was excellent; nevertheless, there were those who proved by arguments which semi-politico-military gossips have always at hand, that the expedition would miserably fail. It was from easy chairs and comfortable rooms in distant cities and cantonments that such inauspicious omens and prophecies of approaching disaster came.

[1] Despatch dated November 2nd, 1824.

1824 November

In November, Máhá Bandula, recalled from Árákán, concentrated at Donabyo an army of 50,000 men well supplied with artillery and entrenching tools. Supplies and reinforcements were also on their way to join the British army; for Sir Thomas Munro, knowing that nothing is so expensive as war carried on with inadequate means,[1] had succeeded in raising the Madras native regiments in Burmah to 100 men above their establishment.

Although the enemy were closing upon the position at Rangoon, their vigilance prevented any accurate information from reaching the general. It was from an intercepted despatch to Martaban that the extent of their preparations became known. It was necessary to look out carefully for their direction and aim. On the 30th of November, from the Shwe-da-gon pagoda, the smoke of their bivouacs could be seen, rising up on the north, east, and west; and in the dense forest, the silence of the night was filled with the crash of falling trees, and the confused sound of multitudes not far off.

December

With early dawn a heavy cannonade and musketry fire told that the post at Kemendine, where Major C. W. Yates commanded, was attacked. His garrison was a detachment of the Madras European Regiment, the 3rd Madras N.I., and a detail of Madras Artillery under Lieutenant Aldritt; and he was supported by part of the fleet on the river under Captain Ryves, H.M.S. *Sophie*. The fighting that ensued was very severe. Again and again during the day the attack was renewed, showing what importance was attached to its possession. The men could not lay aside their arms even while snatching a brief meal. At nightfall a short respite was allowed; but later on in the darkness, as they

[1] Minute dated 3rd August, 1824.

1824 December

prepared to take some well-earned repose, a number of large bamboo fire-rafts, covered with rows of earthen pots filled with burning petroleum, were sent down. Far on either side of the river the reflected light showed the outlines of the forest trees, and the heavy roll of musketry and frequent reports of guns told the anxious watchers in the pagoda that the contest was renewed. But the utmost efforts of the Burmese again failed; and on the river the sailors in their boats had grappled with the danger that menaced the ships, and towed the rafts ashore, there to burn themselves harmlessly out.

In the forenoon of the same day, some 10,000 of the enemy debouched on the plains of Dalla, where they commenced erecting batteries against the shipping, and opened a fruitless fire. Still later in the day, several columns issued from the forest to the east front of the pagoda, and occupied the whole range of hills from thence to Pazandun. The whole of the British position was thus invested on the land side.

Owing to the indisposition of Lieutenant-Colonel Hopkinson, Captain Murray commanded the artillery and superintended all the guns in the lines; Captains Timbrell and Montgomerie were at the pagoda, and Captain Russell was in the town of Rangoon and its vicinity.* Captain John Cheape, Bengal Engineers, then lately joined, was the senior in that service.

The enemy had no sooner taken up their position than they began to entrench themselves in a line of riflepits, each capable of holding two men; and in a marvellously short time all outside of the jungle were under cover. Here and there a gilt umbrella moving

* Captains Murray and Russell had only returned from Pegu on the 1st of December. They had been sent up the river on the 26th of November, with a force commanded by Lieutenant-Colonel J. W. Mallett, of the 89th Regiment.

1824 December

about, showed where some chief was directing operations. Sir A. Campbell, after waiting till they had completed their preparations and brought forward their left into the open ground in front of the White House picquet, where he could manœuvre better, ordered Major Sale, with 400 of the 13th Regiment and the 18th Madras N.I., and a 6-pounder gun, to attack. Twice they advanced, covered by a few rounds from their single gun, and overthrew their opponents; they pierced their second line, and driving them from their entrenchments behind the lakes, returned laden with spears, horse-hair banners, and other barbaric spoil.

Two companies of the 38th, under Captain Piper, in like manner drove back a body of the enemy who had come up too close to the pagoda. The conduct of this officer and his men is particularly dwelt upon in the general's despatch.

For seven days this fighting was incessant. The odds of 50,000 against some 4000 * men, nearly every one but lately recovered from illness, were too great for a first repulse to be sufficiently felt. The fire of fifty guns in position, although the practice was declared by the general to be inimitable, did not prevent the enemy from pushing forward their entrenchments. The post at the great pagoda, commanded by Major Frith, 38th Regiment, suffered severely, as the nature of the ground enabled the enemy to carry on their works to within a very short distance. On the 4th they seemed to have completed their entrenchments on the left, and brought forward their guns and stores from the cover of the jungle. Consequently two columns of attack were formed for the 5th—one of 1100 men, led by Major Sale, and

* Havelock (p. 172) makes the odds 50,000 to 3000; but the above would seem nearer the mark by other accounts.

another of 600 men, by Major J. Walker, 3rd Madras N.I. Captain Chads went with a portion of the fleet up Pazandun creek during the night, and opened the proceedings at daylight. Then the two columns advanced. That on the right, under Major Walker, deployed a little beyond the White House picquet; but its leader, "one of India's best and bravest soldiers,"[1] fell at this moment, and Major Wahab, of the same regiment, at its head charged and carried the entrenchments before him. Major Sale was equally successful, and encamped on the ground he had won a second time. His position near the edge of the large lake east of the pagoda enabled him to menace the enemy's left centre and rear during the two following days.

1824 December

Next morning, the 6th, the artillery fire from the outlying picquets was suspended, but the pagoda and stockade at Kemendine continued to be closely invested. Fire-rafts still from time to time came down the river; everywhere, both seamen and soldiers had to keep unceasing watch throughout the whole of this week of fighting. The duties of their guns kept Captain Timbrell (whose health was suffering severely at this time) and his men employed both day and night, and they repeatedly received the acknowledgements of the general, whose head-quarters were for the present at the pagoda.

On the 7th a final attack was made upon the enemy's position. Shortly before twelve o'clock every gun that could bear upon them opened fire, and four columns commanded by Lieutenant-Colonels J. W. Mallett, B. B. Parlby (30th Madras N.I.), J. Brodie (28th Madras N.I.), and Captain Wilson (38th Regiment), advanced. Major Sale had broken up his camp, and was already acting upon the enemy's communications. There was no check;

[1] Despatch dated 8th December.

1824 December

the Burmese were driven from their entrenchments at all points, abandoning the whole of their guns, swivels, and *matériel*, including the scaling-ladders prepared for an assault on the pagoda.

The loss of the artillery in these seven days was comparatively small:—Bengal, 3 rank and file, 3 lascars and bheestees wounded; Madras—1 lascar killed, 2 rank and file, 4 golandáz, and 1 lascar wounded. The general, in his despatch of the 8th, thus notices them:—

> "The services of the artillery from the three presidencies, commanded by Captains Timbrell and Montgomerie, under the general direction of Captain Murray in the lines, and by Captain Russell, of the Bombay Artillery, in the town and its vicinity, were most conspicuously brilliant." *

And Major Yates, in reporting the attack on Kemendine, speaks of the services of Lieutenant Aldritt and his men in the highest terms.†

Next day, a detachment under Major C. Ferrier, 43rd Madras N.I., subsequently reinforced by Lieutenant-Colonel Parlby, attacked and drove the enemy from the town of Dalla. Captain Russell commanded the detail of artillery employed on this occasion.

Máhá Bandula had not however retreated far. He rallied his scattered men at a rising ground near Kokkeing. The stockades here were more formidable than any which had been attacked, and were held by more than 20,000 of the enemy. On the morning of the 15th Sir Archibald moved out with about 1500 men; Captain Montgomerie commanded the artillery. A column under command of Brigadier-General Willoughby Cotton, 47th Regiment, who had joined the army two

* The names of these officers are again honourably mentioned in G. G. O. Secret Department, dated Dec. 24th, 1824.

† The really valuable services of this officer would not be fairly illustrated by quoting the extract, which is too hyperbolical in expression. *Vide* "East India Military Calendar," vol. iii. p. 298.

1824 December

days before, was sent round to assail the position in rear. His force consisted of the 13th Regiment and 300 of the 34th Madras N.I. Some field-pieces manned by Bengal gunners were attached, but the names of the officers are not known. A squadron of the body-guard, under Lieutenants E. C. Archbold, and E. F. O'Hanlon, officiating adjutant of the Bengal Artillery, accompanied—the last as a volunteer. As soon as General Cotton had arrived in the front of the position on that side, he fired three signal guns, as agreed upon, and forthwith ordered his men on to the attack. Major Sale leading, was severely wounded in the head, and Major Dennie went on with the 13th. They carried one defence after another; but they were outnumbered, and their position became very critical. Their rear was assailed by the enemy's horse, but a gallant charge of the body-guard overthrew these and saved the column. In this charge, Lieutenant O'Hanlon, a very promising officer, was mortally wounded, shot through both arms and in the breast. He died next day. Meanwhile, the rest of the force, in two divisions, under Lieutenant-Colonel E. Miles, C.B. (89th), and Major Evans (38th Regiment), stormed, under a fire from their guns, manned by the Madras gunners under Captain Montgomerie. The place was carried, but with heavy loss. In Cotton's column, 3 European and 1 native officers, and 11 men were killed; 10 European and 2 native officers, and 56 men wounded. Of these the 13th Regiment alone lost 3 officers and 9 men killed, 8 officers and 42 men wounded. The casualties in the rest of the attacking force were 3 men killed, 4 officers and 41 men wounded.

The Bengal Artillery had an officer and 4 lascars wounded, the first mortally; the Madras Artillery had 1 lascar wounded.

1824 December

Rangoon was now free from investment, and Sir Archibald was at liberty to arrange his long-looked-for advance into the kingdom of Ava.

The health of the troops had considerably improved; reinforcements also were arriving, or on their way to join. "From Bengal came two corps, which excited great hopes, and never disappointed them"[1]—the rocket troop under Captain C. Graham, and the 1st Troop, 1st Brigade Horse Artillery under Captain Thomas Lumsden. The first arrived in the end of December. With it were Lieutenant James Paton, 2nd Lieutenant Geo. Campbell, and Sub-Lieutenant Allen.* A part of it remained behind in Bengal. Lieutenant-Colonel George Pollock and Captain E. Biddulph joined the army at the same time. The former had been ordered home for his health, but volunteered for service in Burmah. Previous to leaving Calcutta, he had obtained the permission of the commander-in-chief to arrange with the principal commissary of ordnance at Fort William, Lieutenant-Colonel G. Swiney, for the supply of the stores which he knew the army stood in urgent need of, and immediately on reaching Rangoon he set to work, entering into the minutest detail of his command. Ammunition wagons, with which the Bengal Artillery—probably under the idea that the country was not adapted for wheeled transport—had not been furnished, were supplied ready packed for service;

B-F R.H.A.

A-C R.H.A.

* This officer was formerly in the Royal Navy. In 1816 he was sent out by the Court of Directors with Mr. Wavel, in charge of the first consignment of Congreve rockets, for the purpose of instructing the rocket troop in the use of that weapon. They were appointed to the rank of deputy commissary and conductor of ordnance respectively (G. G. O. 13th Sept., 1816), and posted to the rocket troop. Mr. Allen subsequently had the rank of sub-lieutenant conferred upon him, and was transferred permanently to the Company's service. He retired a few years after this war on a pension.

[1] Havelock.

bullocks were largely imported * or purchased, and a corps of drivers organized from among the syces of the body-guard and horse artillery, as well as the natives of the country.

1824 December

Captain Lumsden, with the 1st Troop, 1st Brigade Horse Artillery, left Calcutta early in January, 1825, and in a few days reached Rangoon. His subalterns were Lieutenants Alexander Thompson, Henry Timings, and Charles Grant.

January

Other changes had taken place in the Bengal Artillery. Lieutenant G. S. Lawrenson had been obliged to obtain temporary sick leave, and Lieutenant O'Hanlon acted for him. On the death of the latter, Lieutenant Errol Blake officiated until Lawrenson's return.[1] The health of Captain Timbrell had suffered so much from long-continued dysentery that he was obliged to return to Dumdum. He had been one of Sir Archibald Campbell's most trusted subordinates, and fully returned that confidence, as indeed did all who served under that officer. It will not detract from the merits of the other artillery officers to mention that the general declared that "Captain Timbrell's boys were always ready for work, and knew how to do it." Official records mention more than once with commendation the names of Captains Russell, Murray, Kennan, and Dickenson, of Lieutenants Seton, Aldritt, Symes, and Onslow; and both despatches and private letters unanimously testify to the

* The want of draught cattle had previously been much felt; but in the position of the force at Rangoon, hemmed in as they were, the country yielded nothing. Sir Thomas Munro had collected all that the army had received. Madras altogether furnished 3500 head of draught bullocks during the war.—"Life of Sir T. Munro;" letter to Lord Amhurst, dated 2nd Feb., 1825; letter to Duke of Wellington, dated 16th April, 1826.

[1] Expeditionary General Orders, December 20th, 1824—G. O. C. C. January 6th, 1825.

1825 January

value of Captain Montgomerie's services and energy as an artillery officer.

Two more squadrons of the body-guard in December, the 47th Regiment in January, and the 2nd Battalion 1st Royals,* completed the reinforcements which the army received. It was a strange mistake to leave the army so unprovided with cavalry. Of this arm there was only the body-guard, numbering 253 sabres, which accompanied Sir Archibald Campbell.

February

Previous to his advance from Rangoon, the general detached Lieutenant-Colonel Godwin with a force to clear the Leing river. He proceeded up the stream on the 6th February, on board the *Satellite*, in tow of the *Diana* steamer, and reached Tantabeng at five in the evening. The enemy opened fire, which was replied to at half-musket range by the guns of the *Satellite*, supported by a shower of rockets from Captain Graham, on board the steamer. Under this the grenadiers landed, and in ten minutes the place was stormed and taken. Thirty-four † out of thirty-six guns were captured. Captain Chads, R.N., commanded the naval part of the force.

For the advance two columns were formed. One, under Sir A. Campbell, to proceed by land, marched on the 13th, about 2400 strong. Colonel Hopkinson commanded the artillery with it, consisting of the rocket troop, with Captain C. Graham, 2nd Lieutenant G. Campbell, and Sub-Lieutenant Allen; the horse artillery, with Captain Lumsden, Lieutenants Timings and Grant. The other column, 1169 strong, under Brigadier-General W. Cotton, was to proceed by water to Tharáwah, and, after taking the enemy's entrenched positions at Panlang and Donabyo on the way, to unite with the

* Royal Scots. † The returns only account for 27.

commander-in-chief. The flotilla consisted of 62 boats, carrying each one or two pieces of ordnance, and the boats of all the ships of war off Rangoon, under command of Captain T. Alexander, C.B., R.N. It sailed on the 16th. The artillery with it were a part of the rocket troop, under Lieutenant Paton, and some foot artillery, under Lieutenants Counsell and Macgregor; Captain Kennan, Lieutenants Syme and Onslow.

1825 February

A force of 780 men was also detached to Bassein, under Major Sale, with orders—after occupation of that town, which was reported to manifest a friendly disposition towards the British—to cross the country to Henzádeh, on the Irawádi, and join the main army. It sailed on the 17th. Only 13 gunners, under a non-commissioned officer, went with it.

Sir A. Campbell reached Mophi on the 17th, Leing on the 22nd of February, and Tharáwah, on the Irawádi, on the 2nd of March, where he halted till the 7th, to obtain news of General Cotton's force before moving on to Prome, his first objective point. But the town of Tharáwah was deserted by its large population, and the few natives who were found were too much intimidated by the cruel policy of the Burmese to afford much reliable information. On the 7th the sound of a distant cannonade came up the river from early morning till nearly two in the afternoon. The reports of some natives told of Bandula's defeat. The general, therefore, pushed on to Thek-kay-beng on the 9th, and to Yádit on the 10th. Here he received the intelligence that Cotton had failed at Donabyo, and, immediately countermarching, he reached Tharáwah on the 13th. Crossing the broad stream of the Irawádi with only about fifteen canoes was tedious work. Rafts were constructed for the ordnance and commissariat stores. Head-quarters,

March

1852 March meanwhile, were fixed at Henzádeh. On the 21st the army resumed its march, and reached Donabyo on the 25th.

February General Cotton, proceeding up the river, had anchored on the evening of the 18th of February near Panlang. Here were two stockades, one on each side of the river, which served as outworks to the main one higher up, where two streams met. Landing his artillery on the 19th, General Cotton directed them to construct a battery on a point of land about 500 yards from the nearest stockades; and in an hour after the order was given, Captain Kennan and his subalterns opened fire from four mortars and two 6-pounders. Lieutenant Paton rendered essential service with his rockets from the deck of the steamboat, anchored midway between both outworks. All three positions were abandoned ere the attacking columns had reached them. Captain Kennan and Lieutenant Paton, as well as Lieutenants Onslow and Symes, were honourably mentioned in the general's despatch.[1]

Considerable difficulty had been experienced in getting the heavier boats up the river, and the *Satellite*, which had grounded, somewhat delayed these operations. The detachment of the 18th Madras N.I. was left at Panlang to secure the communications, and the general proceeded

March with the remainder. On the evening of the 6th of March he was before Donabyo.

Bandula's position consisted of three stockades on the right bank of the Irawádi. The principal one, highest up the stream, was about 1000 yards in length, and from 500 to 800 in breadth, and was situated upon the high bank washed in the monsoon by the swollen river. It commanded two minor works, one at a distance of

[1] Dated Panlang, 24th February, 1825.

500 yards, and the other 400 yards further down. This last was about 200 yards square, and enclosed a pagoda overlaid with white and polished cement, which, glittering in the sun, was visible to the army ten miles. The stockades were constructed of heavy teak beams from fifteen to seventeen feet high, with platforms, and pierced for ordnance and swivels. Outside, a deep ditch with *trous-de-loup*, then several rows of railings, and lastly an *abatis*, completed the defences. The garrison did not fall short of 12,000 men. Held by Máhá Bandula, therefore, it was sufficiently formidable to demand caution on the part of the English general. The breadth of the Irawádi at this place was about 700 yards, so that the Burmese guns covered this line of advance. General Cotton, seeing how the lesser works were commanded by the greater one, was most anxious to attack the latter first. But the commissariat stores for both his own and the land column had been entrusted to his charge; deducting necessary guards and twenty-five sick, he had only 600 Europeans besides his artillerymen; and the naval commander, Captain Alexander, on being asked, said that half these would be required to keep open the river communications. So he abandoned the intention. At sunrise two columns under Lieutenant-Colonel O'Donoghue (47th) and Major Basden (89th Regiment), 250 bayonets each, were disembarked a mile below the pagoda. Captain Kennan with two 6-pounders, and Lieutenant Paton with rockets, covered the advance. The smaller work was carried, and the garrison driven out with great slaughter; but the attack on the centre stockade failed. Two more 6-pounders, four 5½-inch mortars, and rockets were brought up; and after a short bombardment, a storming party of 200 men under Captain Rose, 89th Regiment, was sent forward.

1825 March

The defences were very strong, and they suffered terribly. Captains Rose and Cannon were killed. The attempt was hopeless, and the survivors were recalled. Two 8-inch mortars and four light 12-pounders from the gun-boats were brought up; but towards the close of the day General Cotton gave it up, after losing 128 officers and men in killed and wounded. The Bengal Artillery had one man killed and one wounded; the Madras Artillery six, including a lascar, wounded.

The general retired to Young-young, his encampment of the day before, and briefly reported to his chief the failure and its cause. Sir Archibald's hope of reaching Ava this time had to be deferred.

He had not, as he thought, turned the enemy's position; Bandula's communications were still untouched. The two forces united below Donabyo, heavy guns and mortars were landed, and, the place being too extensive for a complete investment, trenches were opened on both banks of the river against its southern side. Several sorties were made by the enemy, both by night and day. "The energy and activity of the Bengal Horse Artillery and Rocket Troop, under Captains Lumsden and Graham, in assisting to repulse one of these," and "the unremitting zeal of Lieutenant-Colonel Hopkinson and Captain A. Grant, commanding the artillery and engineers," in carrying on the approaches, are mentioned by Sir Archibald in his despatch. On the 1st of April, the rocket and mortar batteries on the right bank opened from a distance of nearly 400 yards; and early next morning the breaching guns had only just commenced, when the report was brought from the right, that a body of the enemy was retiring through the jungle to the westward. The stockades, so active the night before, were silent on that morning. They were already deserted, a fact first ascer-

April

tained by two officers of engineers.* The enemy had lost more than their position: they had lost the only commander who, throughout the war, showed any enterprise or talent for fighting. Máhá Bandula had been mortally wounded the day before, it was said, by a rocket, and no efforts of the other chiefs could allay the panic among their men. All their ordnance and stores were left within the works: 139 guns, from 24 to 1-pounders, and 269 jingals. Among the casualties were: Horse Artillery, one gunner, three natives killed; four natives, two horses wounded. Foot Artillery, Lieutenant G. F. Symes and two gunners wounded, the first severely.

1825 April

The general pushed on at once to Prome, which was reached without any opposition on the 25th. But too much time had already been lost, and here the army had to halt during the rains. Rangoon was now repeopled, and the natives of Pegu showed their gratitude to "the Ingli Rájás, who paid for everything they took, and did not cut off their heads," by restoring plenty to its bazars. The rains, however, which commenced in May closed the land communications of the army, which now depended on the river for all supplies from the seaboard.

Of the troops left at Rangoon on the first advance of the two columns, the Royal Scots and 28th Madras N.I. had reached Tharáwah on the 8th April. Lieutenant-Colonel G. Pollock had joined at Donabyo, and thenceforward remained in command of the Bengal Artillery; Lieutenant-Colonel Kelly, with the Madras European Regiment, occupied Donabyo; Brigadier Smelt commanded at Rangoon, Lieutenant-Colonel Smith at Martaban, and Major Frith at Mergui.

* Most accounts ascribe the discovery to two lascars of the artillery. But a marginal note in pencil has long existed in the copy of Snodgrass's Narrative in the Bengal Artillery Regimental Library, which confirms the above statement.

1825 April — When the head of the British column reached Prome, it was in flames. The enemy had resolved on making a stand there, but the rapid advance, in spite of proffered negotiations, defeated their object of concentration, and they had to retire. About 100 pieces of ordnance were mounted upon the works, and extensive granaries were seized. Ruined villages and wasted crops marked the retreat of the Prince Tháráwádi, the Burmese commander. A reconnaissance was pushed up the river in boats as far as Meaday, sixty miles north of Prome, on the 27th of April; and on the 5th of May Lieutenant-Colonel Godwin, with a force, reached the same place by land, but found no enemy. Two horse artillery guns accompanied him.

May

The two following months passed away quietly; but the enemy were again collecting, and in August a force had occupied Meaday. Meng Myaboo, a half-brother of the king, now commanded. The general, without suspending his preparations for an advance upon the capital, threw up entrenchments along the whole of his front. Another reconnoitring expedition of fifty men of the Royal Scots sent up the river, under Brigadier-General Cotton, found the enemy entrenched and in force. More fruitless negotiations were terminated by a fruitless interview on the 2nd October; and on the 1st November the armistice, which had been agreed to, was concluded by the reception of a distinct refusal to comply with the demands of the British Government.

August

Sept. Oct.

November

During the interval some changes had taken place among the artillery officers. Lieutenant A. Thomson had died at Prome, on the 11th May. A present state of the artillery is given.* The health of the troops appears to have been good. Lieutenant Timings, who had been

* Note G in the appendix to this chapter.

sent back to Dumdum for his health, returned in October with a draft of men and horses; 2nd Lieutenants G. T. Graham, J. H. Daniel, A. P. Begbie, and J. Brady also joined the army with him. Lieutenant P. J. Begbie, of the Madras Artillery, also joined about this time; and a little later, Lieutenants R. C. Moore and F. Burgoyne, of the same regiment.

Meanwhile, the enemy had closely blockaded Prome in front and rear. The ordnance bullock-drivers, who had been largely supplied from the Pegu districts, dreaded a repetition of the cruel barbarities practised on them by the Burmese and deserted in numbers, and but 20 were left out of 180. Forage for the cattle was indispensable, but the British force had no cavalry except a few dragoons of the body-guard. An unhappily devised expedition was sent on the 15th—16th November against a central post of the enemy at Wáh-htee-gán, 13 miles distant, in a north-east direction, but further by road. It failed from the impossibility of securing the simultaneous co-operation of three columns moving by circuitous routes against a distant point over a country covered with thickets and swamps. The loss of two officers and 152 men wounded and missing was the penalty for this error. It had, however, the effect of rendering the enemy boastful and over-confident. In the third week in November 49,000 men were before Prome. Their right rested on the Irawádi at Napádi, the white pagoda of which could be seen from the English lines. The centre stockades were still at the distant post of Wáh-htee-gán, but the left was thrown forward to Tsenbike on the Nga-weng (Nawaing) river.

But though the British general had covered his front with entrenchments, he did not intend to remain behind them. Want of forage alone would have compelled him

1825 November to attack; he determined to fall upon the enemy's left. Two divisions were formed, the first led by himself, the second commanded by General Cotton. The 1st Brigade, 13th and 38th Regiments, was under Lieutenant-Colonel Sale; the 2nd Brigade, 47th Regiment and 38th Madras N.I., under Lieutenant-Colonel R. G. Elrington, of the former corps; the 87th Regiment, with its memories of Barosa, had just arrived, and was strong enough to form a demi-brigade of itself. Lieutenant-Colonel Hopkinson commanded the artillery with this division, to which the horse artillery and rocket troop were also attached.

General Cotton's division consisted of a single brigade, composed of the 41st and 89th Regiments, and the 18th and 29th M.N.I. Lieutenant-Colonel Pollock commanded the artillery, Captain Biddulph's company, with four 8-inch mortars, four 5½-inch howitzers, and three 6-pounder guns. Four native regiments guarded the entrenchments at Prome, while Commodore Sir James Brisbane with the flotilla and the 26th Madras N.I. was to effect a diversion by a feigned attack along the river line.

As the works at Tsenbike were said to cover both sides of the Nga-weng, General Cotton was ordered to move up the left bank, while Sir Archibald Campbell, crossing at the ford of Ziouke, marched along the right. The night of the 30th of November was a busy one for the artillery, who had to withdraw from the works round Prome the ordnance intended to be taken with the attacking force. Before daylight on the morning of the 1st December it moved out, and had reached the ford, when the roll of a heavy cannonade from the river announced Sir James Brisbane's well-timed co-operation. The 1st of Division met with obstacles, and was delayed. General Cotton had made a rapid march of upwards of

December

three hours, and thought of halting a few minutes to refresh his men; but Colonel Pollock was for going on. They soon came in sight of the stockades, covered by the fire of two 5½-inch howitzers under Captain Biddulph. Lieutenant-Colonel Godwin led the attack. A stout resistance was made, and the female leaders of the Shán* troops were to be seen urging on their men. But the contest, though bloody, was short. The Burmese leader, Máhá Nemyo, who commanded at this point, was found among the dead. The 1st Division was too late to share in the fight, and returned to Ziouke, whither Brigadier Elrington had been detached to cover the communications with Prome. The 2nd Division bivouacked at Tsenbike.

Next morning Sir Archibald attacked the enemy's right at Napádi. The first brigade of the 1st Division advanced against the pagoda hill along the river bank. The horse artillery and rocket troop were here. The 87th was ordered to attack the centre; the second operated against the left centre and left; while the 2nd Division, Lieutenant-Colonel Pollock commanding the artillery as before, was to penetrate the woods to the British right, and turn the enemy's flank in that direction.

Captain Lumsden, with his troop, was sent to take up a position in advance. Over rocks and down the beds of nullahs he made his way, but a steep ridge prevented his getting within efficient range. Captain Graham with the rockets supported Lumsden. An unlucky accident, however, had nearly deprived the regiment of the latter gallant officer. A howitzer missed fire twice, and Lumsden, standing by, ordered the shell to be withdrawn. But though the charge had not ignited, the fuze had, and a lascar was killed, one of the gunners badly

* A race from the north-eastern portion of the Burman empire, fairer and taller than the Burmese.

1825 December

wounded, and Captain Lumsden thrown down, severely scorched and contused. He sprang up, however, to the surprise of all, and, seating himself, continued calmly to direct the fire of his guns. His name was not even returned among the roll of wounded.

Meanwhile, the two brigades of the 1st Division and the 87th were busily engaged. The 38th led the attack of the first brigade, turned an *abatis* at the foot of the hill, and ascended; the 87th, disdaining a circuitous route, forced a passage through the strongest of the advanced defences, and joined it in the assault. Elrington fought for every foot of ground, and was not so far in advance; Cotton was beaten by the thick high grass jungle, and was obliged to send the guns back; Lieut.-Colonel Pollock and his men had to bivouac upon the plain.

The defeat of the enemy was most complete; seventeen iron and three brass guns, one of the latter a 32-pounder, were taken. The casualties in the artillery were all in the 1st Troop, 1st Brigade R.A.; one lascar killed and eleven gunners wounded appear in the returns. Lieutenant G. A. Underwood (Madras) and Lieutenant F. Abbott (Bengal Engineers) were also wounded. Sir Archibald, in his despatch of the 4th, wrote thus:—

"Lieutenant-Colonel Hopkinson, commanding the artillery, Lieutenant-Colonel Pollock, and Captain Graham, of the Bengal Artillery, merit my fullest approbation for their exertions; and Captain Lumsden, of the Bengal Horse Artillery, although badly wounded, refused to quit the battery, and continued from his chair to direct the fire of his guns."

There only now remained to dislodge the portion of the Burmese force across the river. This was effected by General Cotton on the morning of the 5th, with details from the 1st, 41st, and 89th Regiments, a company of native infantry, and some pioneers. Lieutenant Paton with a division of the rocket troop, and Lieutenant R. S. Seton (Madras Artillery), with four howitzers, accompanied, and were mentioned in despatches.

1825 December

The army moved forward by different routes, and on the 19th was at Meaday. Here the number of crucified peasants and half-buried Burmese soldiers told of the ravages war and cholera were making in the enemy's ranks, and of the cruelties by which the aid of the villagers was purchased by them. The unfortunate natives tendered their services to the British with readiness, notwithstanding the fate that might hang over them. The artillery drivers, composed of Burmese,* Madras pioneers, syces, grass-cutters, and gun lascars, presented a motley appearance. On the 29th the army reached Patanagoh, on the Irawádi, opposite to Melloon, where the enemy were making another stand. Sir James Brisbane, with the flotilla, had come up the river, passing a very strongly fortified post at Palho (or Sak-ka-doung), which was deserted. On nearing Melloon, he determined to run the gauntlet and pass the stockades. To the astonishment of the army looking on, and doubtless to that of the gallant commodore himself, the boats passed silently along without receiving or returning a shot, and took up their ground beyond. The river was not more than 500 yards† wide here. The Burmese position formed a quadrangle, the interior of which was too much exposed to the left bank. In the centre was a small hill surmounted by a pagoda, and fortified. Outside, close

* Fifty of these, taken from the Irawádi boatmen and sent to Colonel Pollock, brought each man his oar with him.—"Diary of Sir G. Pollock," December 6th.

† So Havelock says. Captain Trant says 1000; but from the efficient practice of the horse artillery in enfilade, the former must be nearer the mark. It must be remembered at the same time that the ordnance of the troop was "mixed," according to the old equipment, and contained two 12-pounders.

1825 December to the river bank at the south-east angle, was a new pagoda, freshly gilt all over—the cenotaph erected by the king's order to Máhá Bandula.

But negotiations once more commenced. A treaty even was drawn up, signed, and was reported to have been sent to Ava for the ratification of the sovereign. So much did general expectation point towards a final cessation of hostilities, that Captains Graham and Biddulph and Lieutenant Paton obtained leave, and on the 5th of January started for Rangoon by water. 1826 January "The hopes of Ava" were, however, not yet broken, for the truth had not reached the ears of the king.

There was a force at this time stationed in Pegu under command of Lieutenant-Colonel H. H. Pepper, of the Madras service. Captain Dickenson commanded a small detail of artillery with it. The Burmese in this quarter having become troublesome, Lieutenant-Colonel Pepper had marched to Shwe-gein, which he occupied on the 3rd of January without opposition. From thence he detached Lieutenant-Colonel Conry to reduce Sitang, a stockaded post between Tonghoo and Martaban. But it was too strong for the small force sent; Lieutenant-Colonel Conry was killed, and it was repulsed with loss. Pepper, therefore, stood before the place on the 11th. It was of considerable extent, situated on an eminence; the teak walls, from twelve to fourteen feet high, commanded every approach. A creek of the river, fordable only at low water, covered the northern face. While preparations for the assault were being made, Captain Dickenson,* with a

* Major Begbie, in the "Services of the Madras Artillery," mentions Lieutenant J. C. Patterson as having been present on this occasion (vol. ii. Appendix No. 1, p. xxviii.), but this officer is not mentioned in the official return ("Documents," p. 203), which gives one captain and sixteen gunners as the party sent on this service. It is possible that Lieutenant Patterson may have been present, though otherwise employed.

6-pounder gun and 4⅖-inch howitzer, kept up a fire which mainly contributed to keep down that of the enemy. The place was taken, but with severe loss to the assailants. Three hundred dead bodies were counted within the enclosure, besides many others thrown into wells or carried off. Captain Dickenson was honourably mentioned in Lieutenant-Colonel Pepper's despatch of the 14th. After this Pegu remained quiet.

1826 January

The 18th of January was the last day given by Sir Archibald for the production of the ratified treaty, which was still lying unsigned in the tent of Meng-Myaboo.

At midnight the landing of the ordnance commenced, and busy parties carried to the appointed places the materials for the batteries.

"His lordship in council," writes the general, "will be enabled to appreciate the zeal and exertions with which my orders were carried into effect, under the direction of Lieutenant-Colonel Hopkinson, commanding the artillery, and Lieutenant Underwood, the chief engineer (aided by that indefatigable corps, the 1st Battalion of Madras Pioneers, under the command of Captain Crowe), when I state that by ten o'clock next morning I had 28 pieces in battery, on points presenting a front of more than a mile on the eastern bank of the Irawádi."[1] The guns and howitzers of the horse artillery were placed opposite, to the left of the central work.

At 11 a.m. Sir Archibald gave the signal, and peal after peal reverberated through the rocks and woods round Melloon. Anxious eyes watched the effects of the fire. "It was evident," says the deputy assistant adjutant-general, "that the artillerists had hit the range

[1] Sir A. Campbell to Secretary to Government, Secret Department, dated January 20th, 1826; Documents, etc.; Appendix, p. 194.

1826 January

at once. Balls were seen to strike the works, raising a cloud of dust and splinters, demolishing the defences, and ploughing up the area of the square. Shells lit sometimes a few paces from the parapet behind which the garrison was crouching, bursting among their ranks, sometimes upon the huts of the troops and marked points of the pagodas. The rockets flew in the truest path. . . . . Twice the line of the barbarians, which manned the eastern face, gave way under the dreadful fire; twice they were rallied by their chiefs." [1]

This bombardment lasted an hour and a quarter before the attacking party commenced to move across the river. The 1st Brigade, as the stream carried the boats down past the stockades, suffered very much. Colonel Sale was left severely wounded in one of them, and Major Frith, 38th, led the assault. The 13th and 38th together, now numbering only 480 bayonets, as usual contended for the foremost place. General Cotton, with three columns under Brigadiers H. Godwin, B. B. Parlby, and Hunter Blair, was ordered to cross above the works and attack the northern face, but the co-operation was not timed sufficiently accurately to answer the proposed end as fully as was intended. The 1st Brigade, under Major Thornhill, of the 13th, Major Frith having been severely wounded, had already driven out the enemy. Seventy-four pieces of ordnance, brass and iron, were captured.

In his despatch the general notices the service rendered by the artillery in the following terms:—

"Where zeal displays itself in every rank, as among the officers whom I have the happiness to command, and all vie with each other in the honourable discharge of duty, the task of selecting individual names for the notice of his lordship becomes difficult and embarrassing, and I am compelled to adopt the principle of particularizing those alone on whom the heaviest share of duty

[1] Havelock, p. 300.

devolved on this occasion. It fell to the lot of the artillery to occupy this conspicuous station in the events of the day. In behalf, therefore, of Lieutenant-Colonel Hopkinson commanding the whole, and of Lieutenant-Colonel Pollock commanding Bengal Artillery, and Captains Lumsden, Bengal Horse Artillery, and Montgomerie, Madras Artillery, commanding the batteries, I have to solicit your recommendation to his lordship's favourable attention. The rocket practice under Lieutenant Blake,* of the Bengal Horse Artillery, was in every way admirable; of 304 rockets which were projected during the day, five only failed of reaching the spot for which they were destined,† and uniformly told in the works or in the ranks of the enemy with an effect which clearly established their claim to be considered a most powerful and formidable weapon of war."

1826 January

A further advance now became necessary. Sir Archibald marched on the 25th, General Cotton following. On the 3rd February the leading column reached Pakang-ye, opposite Sem-byo-gyun, where the road across the hills from Árákán joined. But General Morrison's division had ere that ceased to exist as an army, and no co-operative force was there. The forage was very scarce, and losses among the cattle had considerably increased the difficulty of moving. Of the few troopers of the body-guard, most were mounted on ponies and private animals; officers' chargers were put in requisition "to assist in dragging the guns of the invaluable horse brigade." [1]

February

The Burmese army had now a new commander, who had volunteered to beat his foes in open field. He had assumed the title of "Prince of the Setting Sun," or "of darkness," construed by a not unnatural transition of idea in the minds of the British soldiers into "the King of Hell."

* Lately transferred from 3rd Company, 5th Battalion, to the 2nd Troop, 2nd Brigade, now the designation of the rocket troop.

† These rockets had not then been sufficiently long in India to suffer from the climate.

[1] Havelock, p. 314.

1826 February

On the 9th the general moved from Yassay on Logánanda. The enemy's force at this place was estimated at 16,000 men. To oppose it there were, after deducting a native regiment left to guard the baggage and stores, only these numbers:—

	Men.
Artillery	116
Body-guard	33
13th Light Infantry	216
38th Regiment	281
41st „	249
89th „	148
43rd M.N.I.	251
	1294

The advanced guard was composed of two companies of the 13th, the body-guard, and the horse artillery (four guns). After its coming into collision with the enemy, it was intended that the general with the 13th and 89th should attack the enemy's left, and General Cotton with the other two their right, supported by the foot artillery, while the native regiment preserved the communication with the river. The road was enclosed with a thick jungle of *ber* * bushes, extending for miles around.

The advance drove in the nearest of the enemy posted at Logánanda, while the 13th, merging into the more open ground, spread out *en tirailleur*, and with the guns were soon hotly engaged. A small portion of the army was thus extended over a considerable portion of ground. But the rest of the column, on a road taken up for considerable spaces with ammunition and rocket carriages and guns, could not, from the nature of the ground, debouch sufficiently soon, and General Campbell was in considerable peril. At one time he had with him but fourteen of the 13th, sixteen troopers, and two of the

* *Zizyphus jujuba.*

guns. Pressed upon by a shouting multitude of the enemy, he had to retire to a mound, where he took up his post, calling in the skirmishers by bugle. The coolness of the troopers in covering the movement and the conduct of Subadár-Major Kázi Wali Muhammad in particular, was remarkable. It was not long before the rest of this division, with the rockets and foot artillery, were up. Cotton's active energy, meanwhile, had brought up his regiments on the left; already he was forcing back the enemy's right and threatening their communications. Lieutenant-Colonel Pollock and two of the Madras guns were with him. Sir Archibald again advanced with his division, and the enemy, driven from one post to another among the countless pagodas of Pagahm-myo, for the last time disappeared before the face of the British army.

The conduct of the troops was what it ought to be for British soldiers, no mean standard of merit; and the general, at the close of his orders for the day, worthily said—

"The frequency of their acts of spirited soldiership on the part of his troops renders it difficult for the major-general to vary the terms of his praise; but he offers to every officer and soldier engaged this day the tribute of his thanks, at once with the affection of a commander and the cordiality of a comrade."

Before the army left Pagahm-myo on the 16th, it had learned the fate of the luckless "King of Hell," who, rashly hastening back to his king to excuse his failure and demand fresh troops, was dragged forth from his presence and inhumanly butchered in the streets of Ava. Yet he had done his best.

On the 23rd of February the army had reached Yandabo, and here the treaty of peace, with all the conditions demanded, was finally concluded. A deputation of three officers, Captain Lumsden, Lieutenant H.

1826 February — Havelock, the deputy assistant adjutant-general, and Assistant-Surgeon Knox, of the Madras army, were sent with some formal presents to Ava, where, on the March 1st of March, they were received by the king.

The evacuation of Burmah commenced at Yandabo on the 7th of March, at the end of which month most of the troops were on their way back to their own presidencies; Rangoon itself being, under the terms of the treaty, occupied for some months longer.

A donation of six months' batta was awarded to that portion of the army which had spent more than twelve months in Burmah, and of three months' batta to those corps which had served less than that time. A medal was also given to the native troops.

The Government general order published on the promulgation of the treaty of peace mentions, with special acknowledgment, "the services of the Bengal and Madras Foot Artillery, under Lieutenant-Colonel Hopkinson and Lieutenant-Colonel Pollock, and the Bengal Rocket Troop and Horse Artillery, under Captains Graham and Lumsden." The distinction of Companion of the Bath was conferred upon the two first-named officers (dated December 26th, 1826, *London Gazette* of the 2nd January following); and subsequently, on the accession of her present Most Gracious Majesty, on Captains Graham, Lumsden, Timbrell, and Montgomerie.

During the course of writing this chapter, the mail brought the intelligence to India of the deaths, within a day of one another, of Field-Marshal Sir George Pollock and of General Sir Patrick Montgomerie, then the two senior officers of the Bengal and Madras lists of Royal Artillery. This is not the place to detail the merits of either, nor is it necessary. They are justly known far beyond the limits of the regiments in which they served so long and so well. From Sir George Pollock I have received much valuable information regarding this as well as other campaigns, and I would not omit to record here my debt of gratitude to one whom, though never seen, I have had reason to think of as a valued friend.

A few months previously, Lieutenant-Colonel J. H. Macdonald, the youngest of the artillery officers who landed at Rangoon in May, 1824, had also passed away, at a comparatively early age. He, too, has rendered me valuable aid in this work, which I now regretfully acknowledge.

AUTHORITIES CONSULTED FOR THIS CHAPTER.

1. Documents illustrative of the Burmese War. With an Introductory Sketch by H. H. Wilson, Esq. 1 vol. 4to. Calcutta, 1827.

2. Memoir of the Three Campaigns in Ava. By Henry Havelock, Lieutenant 13th L.I. 1 vol. 8vo. Serampore, 1828.

3. Narrative of the Operations of Major-General Sir A. Campbell's Army. By Major Snodgrass. 1 vol. 8vo. London, 1827.

4. Services of the Madras Artillery. By Major Begbie.

5. Memoir of Sir G. Pollock. *Golden Hours* for 1870.

6. Life of Sir Thomas Munro. By the Rev. G. R. Gleig. 2 vols. 8vo. London, 1830.

7. Life of Sir Henry Lawrence. By Major-General Sir H. Edwardes and H. Merivale, etc. 2 vols. 8vo. London, 1872.

8. Madras Artillery Records.

9. General Orders, *passim*.

10. Letters from Field-Marshal Sir G. Pollock; Generals G. Campbell and B. W. Black; Colonels Lumsden, Timbrell, R. S. Seton, Pillans, Macdonald, and Lawrie (Royal, Bengal, and Madras Artillery).

11. Letter from Lieutenant James Low, Madras Infantry, to Colonel Nicol, Adjutant-General Bengal Army.

APPENDIX.

Note A.—Lieutenants Bedingfield and Burlton.

Note B.—Officers of the Bengal Artillery who served on the Assam frontier in the years 1822-1826.

Note C.—Officers of the Bengal Artillery who served on the Árákán frontier.

Note D.—Officers of the Bengal Artillery who served with the army under command of Major-General Sir A. Campbell.

Note E.—Officers of the Madras Artillery who served with the army under command of Major-General Sir A. Campbell.

Note F.—Detail of troops composing the expedition landed at Rangoon.

Note G.—Present state of the artillery.

Note A.

Lieutenant Philip Bowles Burlton was the youngest son of the late W. Burlton, Esq., of Wykin Hall, Leicestershire, and Donhead Lodge, Wiltshire. The cause of his being sent to Assam, and his melancholy death there, are worthy of a note.

About the time that he arrived in India (1821), it is generally known that Mr. J. S. Buckingham, the editor of a Calcutta newspaper, had given much offence to those in power by the freedom with which he discussed the measures of Government, not then prepared to admit the liberty of the press in regard to public affairs. Major-General Hardwick, the commandant of artillery, was a Conservative in the highest sense of the word, and held strong opinions on this point. His indignation, therefore, was great on finding that a young subaltern had, unconscious of the guilt he was incurring, invited Mr. Buckingham to be his guest on some public occasion at the regimental mess at Dumdum. "The mess must not be contaminated," said the commandant, "or revolutionary ideas instilled into the minds of my officers." Wherefore he proposed certain resolutions for their adoption. His power and influence were great, but some of the younger hands were inclined to assert their right to select their company as long as no exception could be taken to the character and social position of the guest. Lieutenants Burlton and C. H. Wiggens were foremost in asserting the right of choice. They were therefore ordered off, one to Assam, the social antipodes of Dumdum, the other to Agra.

Lieutenant Burlton, in his exile, found ample employment, first in the service of that part of the war with Burmah carried on in Assam, and afterwards in following up the course of the Brahmaputra river, with the object of discovering its source, and solving other geographical questions. For nearly seven years he was associated with Lieutenant Richard Gordon Bedingfield in the discharge of his duties, and at last in the same tragical end. They had gone, for the benefit of their health, to Nanklo, near Gaoháti, which had then lately been fixed upon as a sanatarium in the Kásiya hills. Here, on the 2nd of April, 1829, the house was surrounded by some hundreds of Kásiyas and Gáros, the wild tribes of these mountain-ranges, still imperfectly civilized, and Bedingfield, going out among them to inquire what they wanted, was barbarously murdered. Lieutenant Burlton, with a European writer, named Beauman, and a few sepoys, held the house till next day, when it was set on fire, and both were killed in the endeavour to make good a retreat to Gaoháti.

NOTE B.

Officers of the Bengal Artillery who served on the Assam frontier in the years 1822-1826.

Captain Jonathan Scott.
,, Thomas Timbrell.
,, Charles Smith.*
Lieutenant Edward Huthwaite.
,, Richard G. Bedingfield.
,, Philip B. Burlton.
,, Joseph Turton.
,, Frederick Brind.
,, John T. Lane.

---

* Promoted during the war.

NOTE C.

Officers of the Bengal Artillery who served on the Árákán frontier of Burmah in the years 1824-1826.

Lieutenant-Colonel		Alexander Lindsay.	
Captain	...	Edward Hall.	Died on board ship at the mouth of the Talak river, Jan. 14, 1826.
,,	...	John Rawlins.	
Lieutenant	...	John S. Kirby.	
,,	...	Henry Rutherford.	
,,	...	James W. Scott.	
,,	...	James R. Greene.	Died in Calcutta of fever, Oct. 5, 1825.
,,	...	George Hart Dyke.	
,,	...	John Hotham.	
,,	...	William C. J. Lewin.*	
,,	...	Henry M. Lawrence.*	
,,	...	Samuel W. Fenning.*	
Second Lieutenant		John Fordyce.	
,,		Ambrose Cardew.	
,,		Edmund Buckle.	

NOTE D.

Officers of the Bengal Artillery who served with the army under command of Major-General Sir Archibald Campbell, G.C.B., in the first Burmese war (1824-1826). From despatches and various sources.

Lieutenant-Colonel		George Pollock.	
Captain	...	Charles Graham.	
,,	...	Edward Biddulph.	
,,	...	Thomas Lumsden.	
,,	...	Thomas Timbrell.	Returned to Bengal sick, December, 1824.
Lieutenant	...	William Counsell.	
,,	...	George H. Rawlinson.	
,,	...	Alexander Thompson.	Died at Prome 11th May, 1825
,,	...	Birnie Browne.	Deputy Assistant Quarter-master General
,,	...	Henry Timings.	
,,	...	James Paton.	
,,	...	George S. Lawrenson.	Adjutant.
,,	...	Charles Grant.	

* Promoted during the course of the war.

Lieutenant	...	Errol Blake.	
,,	...	Robert G. Macgregor.	
,,	...	Edward F. O'Hanlon.	Killed at Kok-keing, 16th December, 1824.
,,	...	James H. Macdonald.	
Second Lieutenant		George Campbell.	
,,		George T. Graham.	
,,		James H. Daniel.	
,,		Arthur P. Begbie.	

NOTE E.

Officers of the Madras Artillery who served in the first Burmese war (1824-1826). From despatches, Begbie's "Services of the Madras Artillery," and official records.

Lieutenant-Colonel		Charles Hopkinson.	Joined Sir A. Campbell's force October, 1824.
Major	...	William M. Burton.	Left Sir A. Campbell's force October, 1824.
Captain	...	Andrew S. Murray.	
,,	...	John J. Gamage.	
,,	...	T. Y. B. Kennan.	
,,	...	Patrick Montgomerie.	Brigade-Major artillery, Sir A. Campbell's force.
,,	...	William F. Lewis.	Commissary of Stores, Sir A. Campbell's force. Died at Prome 11th December, 1825.
,,	...	Frederick Bond.	
,,	...	John Lamb.	Brigadier-General Morison's force. Died of fever August, 1825.
,,	...	John Dickenson.*	
,,	...	George F. Symes.*	
,,	...	Richard Somner Seton.*	
Lieutenant	...	John Aldritt.	
,,	...	George Alcock.	
,,	...	Adolphus E. Byam.	
,,	...	John C. Patterson.	
,,	...	George Middlecoat.	Brigadier-General Morison's force.
,,	...	George W. Onslow.	
,,	...	Peter J. Begbie.	Joined August, 1825.
,,	...	Thomas E. Geils.	
,,	...	J. G. B. Bell.	Joined at the end of 1825.
,,	...	Richard C. Moore.	
,,	...	Frederick Burgoyne.	

* Promoted to the rank of captain during the war.

NOTE F.

Detail of troops composing the expedition landed at Rangoon in May, 1824, and such as joined the head-quarters of the army up to the 1st of January, 1825.

Regiments.	Date of arrival at Rangoon.	Number including officers.	Total.	Remarks.
BENGAL TROOPS.				
Det. European Foot Artillery	11th May	360		
H.M.'s 13th Light Infantry	"	727		
" 38th Regiment	"	1,035		
Det. 40th Native Infantry	"	24	...	The corps left at Cheduba.
Rocket Troop	28th Dec.	86		
Gov.-General's Body-guard	4th, 24th, and 26th Dec.	353		
			2,585	
MADRAS TROOPS.				
Detachment Foot Artillery	11th May	556		
H.M.'s 41st Regiment	"	762		
Madras European Regiment	"	433		
1st Battalion Pioneers	"	552		
3rd Native Infantry	"	676		
7th " "	"	695		
12th " "	"	652		
9th " "	"	658		
18th " "	"	609		
34th " "	"	617		
43rd " "	"	711		
H.M.'s 89th Regiment	6th June, 22nd Nov.	1,012		
" 47th "	26th Dec.	177	...	The rest joined in 1825.
26th Native Infantry	1st Oct.	636		
28th " "	1st and 3rd Sept.	832		
30th " "	27th Sept.	613		
			10,191	
BOMBAY TROOPS.				
Detachment Foot Artillery	12th June		69	
Total			12,845	

NOTE G.

Present state of the Artillery serving with the army under command of Brigadier-General Sir Archibald Campbell, K.C.B.

Head-quarters, Prome, 18th August, 1825.

	Present fit for duty.												Sick.			
	Field Officers.	Captains.	Subalterns.	Surgeons.	Assistant Surgeons.	Staff-Sergeants.	Subadars.	Jemadars.	Sergeants Havildars.	Drummers.	Rank and File.	Horses.	Sergeants Havildars.	Drummers.	Rank and File.	Horses.
At Prome { Horse Brigade	...	2	4	...	1	...	...	...	9	3	84	129	1	...	17	7
At Prome { European Foot Artillery ...	2	2	1	...	2	...	...	...	8	2	128	...	...	...	22	0
At Rangoon { European Artillery	...	1	3	...	1	1	...	...	2	...	45	...	...	...	11	0
At Rangoon { Native Artillery	...	...	...	...	...	...	2	6	6	3	49	...	1	2	10	0
Total ...	2	5	8	...	4	1	2	6	25	8	306	129	2	2	60	7

CHAPTER XIV.

SECOND SIEGE OF BHURTPORE—Usurpation of Durjan Sál—Sir David Ochterlony's movement against Bhurtpore is countermanded—Preparations for the siege—Brigading of the army—Horse artillery—Foot artillery and battering train—Bhurtpore invested—Water supply for the ditches cut off—Base of attack laid—First batteries established—Desertion of an artilleryman—Cavalry and horse artillery on the west face—Major Whish's battery—Great mortar and breaching batteries—Breaching and bombarding commenced—Mining commenced—Trenches extended on the right—Overtures for peace—Storming postponed—Colonel Stark and Lieutenant Pennington examine right breach—Explosion in the ammunition depôt—Enemy occupy the ditch—Mine under north-east angle—Mine under long-necked bastion fired—Breach examined—Arrangements for storming—Storm and capture—Casualties in the artillery—Ordnance captured—Officers noticed in orders.

SECOND SIEGE OF BHURTPORE.

MÁHÁRÁJÁ Ranjit Singh, who had successfully defended Bhurtpore in 1805, was succeeded by Baldeo Singh, who died in 1824, leaving his maiden fort to an infant son, Balwant Singh. This child had been shortly before formally acknowledged as the successor by the Resident at Delhi, Sir David Ochterlony. His cousin Durjan Sál, son of a younger brother, did his best to set aside the succession; and finally, in March, 1825, terminated a series of intrigues by murdering Rám Ratan, uncle and guardian to the boy, whose person he seized, and nominated himself regent.

1824

1825

Sir David Ochterlony, than whom none of our Indian

1825

political officers had better opportunities of knowing the native mind, saw in this bold act of usurpation more than an act of individual defiance to British authority. In the peculiar position of the Government, then engaged in a distant and tedious war beyond its frontier, and in its relation with the only state that could boast of having successfully resisted our army, and which was a sort of link between the Máhrátá principalities of Central India and the tribes of the north-west, he understood the value of the maxim, "Aut Cæsar aut nullus," and acted upon it. Finding the remonstrances and demands of Government of no avail, he used the extraordinary powers vested in him, and collected from the Agra and Meerut divisions a force which he placed under Major-General T. Reynell, C.B., commanding the latter, for the reduction of Bhurtpore.

But Government was not prepared for so serious an undertaking. Peremptory orders were sent for the immediate dispersion of the force; and the resignation of Sir David Ochterlony, which followed, was forthwith accepted.* He was succeeded by Sir Charles Metcalfe;

* This reversal of his policy broke the old general's heart. He felt himself disgraced in the eyes of the natives, among whom his name was as a household word, and his influence almost unbounded. He died at Meerut, on the 14th of July. It may be doubted whether the force he assembled was in truth sufficient for the purpose—but it was not on these grounds that the measure was opposed by Government. It was disapproved of *in toto*. Nevertheless, the records of his past services, especially in the conduct of the Nipál war, show that Sir David Ochterlony was not likely to rush blindly into danger, or to attack directly a position which might better have been turned. But it is easy for those who lag behind to fasten the reproach of precipitancy upon those who are in front, particularly when the former are the ones in authority. Some did not hesitate even to pronounce Ochterlony past service; but the memorial he penned in his defence shows no failure of intellect, though, after more than forty-seven years of service in India, and at the age of sixty-eight, his health was on the decline. It is strange that among the many men who have written of distinguished Indian officers, Sir David Ochterlony should never have found a biographer.

1825 and thus the year wore away. But in war, as Sir Thomas Seaton well puts it, time is composed of men's lives. In the months thus gained Durjan Sál increased his stores of ammunition and provision, strengthened his defences, largely recruited his followers, and made engagements with all the readily acquiescent chieftains of Central and North-West India. From the ruler of the Punjab he received assurances of help in a struggle which, perchance, might enable him to gain the authority September he longed to exercise over the Cis-Sutlej Sikhs. The Governor-general, though tardily, was at last convinced that Bhurtpore must be taken, and the stain of twenty years' duration be wiped out.

Once determined upon, the gates of war were thrown open wide, and men and *matériel* poured forth. Lord Combermere had just succeeded Sir Edward Paget as commander-in-chief in India; leaving Calcutta, he ar- December rived at Agra on the 1st of December, and assumed command of the army assembled there and at Muttra, numbering about 30,500 men.* It was brigaded as follows:—

INFANTRY.

1st Division.

Major-General Thomas Reynell, C.B., commanding.

1st Brigade (Brigadier-General J. McCombe, 14th Regiment, commanding).—14th Regiment; 23rd and 63rd N.I.

4th Brigade (Brigadier T. Whitehead, 41st N.I., commanding).—32nd, 41st, and 58th N.I.

5th Brigade (Brigadier R. Patton, C.B., 18th N.I., commanding).—6th, 18th, and 60th N.I.

2nd Division.

Major-General Jasper Nicolls, C.B., commanding.

2nd Brigade (Brigadier-General W. T. Edwards, 14th Regiment, commanding).—59th Regiment; 11th and 51st N.I.

---

* A detail of the strength is given in the Appendix, Note A.

· 3rd Brigade (Brigadier-General J. W. Adams, C.B., 4th extra N.I., commanding).—33rd, 36th, and 37th N.I. 1825 December

6th Brigade (Brigadier C. S. Fagan, 15th N.I., commanding).—15th, 21st, and 35th N.I.

CAVALRY.

Brigadier-General J. W. Sleigh, C.B., 11th Light Dragoons, commanding.

1st Brigade (Brigadier G. H. Murray, C.B., commanding).—16th Lancers; 6th, 8th, and 9th Bengal Cavalry.

2nd Brigade (Brigadier M. Childers, 11th Light Dragoons, commanding).—11th Light Dragoons; 3rd, 4th, and 10th Bengal Cavalry.

ENGINEERS.

Brigadier Thomas Anburey, C.B., commanding.
Lieutenant A. Irvine, Brigade-Major.
Six companies of Sappers. Two companies of Pioneers.

ARTILLERY.*

Brigadier Alex. Macleod, C.B. (commandant of the regiment), commanding.
Captain James Tennant, Assistant Adjutant-General.
Lieutenant John S. Rotton, Deputy Assistant Quarter-master General.
2nd Lieutenant Francis Dashwood, Aide-de-camp.

Horse Artillery.

Brigadier Clements Brown, commanding.
Lieutenant Charles R. Whinfield, Brigade-Major.

With 1st Cavalry Brigade.

Major W. S. Whish, commanding.

2nd Troop, 1st Brigade	Captain Roderick Roberts.	C-C R.H.A.
1st Troop, 2nd Brigade	Captain James C. Hyde.	A-F R.H.A.

With 2nd Cavalry Brigade.

Lieutenant-Colonel Henry Stark, H.A., commanding.

3rd Troop, 2nd Brigade	Captain George Blake.	C-F R.H.A.
1st Troop, 3rd Brigade	Captain Henry J. Wood.	B-C R.H.A.

* The list of officers is given in Note B, and the numerical strength of the horse, foot, and lascars, in Note C in the Appendix.

1825 December

Besides the above, there were attached to the infantry as field artillery:—

With 1st Division.

Lieutenant-Colonel John A. Biggs, commanding.
4th Troop, 3rd Brigade, Native Horse Artillery; Captain Gabriel Napier C. Campbell.
A-8 R.A. 2nd Company, 3rd Battalion, with Light Field Battery; Captain Peter L. Pew.

With 2nd Division.

Lieutenant-Colonel Charles Parker, commanding.
4th Troop, 2nd Brigade, Native Horse Artillery; Captain John J. Farrington.
8-23 R.A. 1st Company, 3rd Battalion, with Light Field Battery; Captain William Curphey.

B-F R.H.A. The portion of the 2nd Troop, 2nd Brigade, which had not proceeded to Burmah, was also present; and as Lieutenant Whinfield, the only remaining officer with it, was on the staff, it was attached to one of the other troops. It had no rockets with it.

D-C R.H.A. Another troop, the 2nd Troop, 3rd Brigade, under the command, temporarily, of Captain William Bell, of the 3rd Troop, was also present; but having only just been raised, the guns and horses were left at Agra, the officers and men joining other troops or the park.

Foot Artillery.

Brigadier R. Hetzler, C.B., commanding.
Lieutenant James Johnson, Brigade Major.
Captain Isaac Pereira, Commissary of Ordnance.
Captain George Brooke, Deputy Commissary of Ordnance.

A-19 R.A.	2nd	Company,	1st	Battalion,	Lieutenant Augustus Abbott.
E-16 R.A.	3rd	,,	,,	,,	Captain George H. Woodroofe.
C-16 R.A.	4th	,,	,,	,,	Lieutenant John Edwards.
8-23 R.A.	1st	,,	3rd	,,	Included with the Horse Artillery.
A-8 R.A.	2nd	,,	,,	,,	
D-16 R.A.	4th	,,	,,	,,	Lieutenant Edward S. A. W. W. Wade.
A-16 R.A.	2nd	,,	4th	,,	Lieutenant Archdale Wilson.
7-23 R.A.	3rd	,,	,,	,,	Lieutenant Rowland C. Dickson.

The ordnance sent with the army was as follows:—

1825 December

FIELD ORDNANCE.

	12-pounder guns.	6-pounder guns.	24-pounder howitzers.
4 European troops H.A. ...	8 ...	8 ...	8
2 Native „ „ ...	0 ...	12 ...	0
2 Light Field Batteries ...	6 ...	6 ...	2
Total ...	14	26	10

SIEGE ORDNANCE.

24-pounder guns	16	
18 „ „	20	
12 „ „	4	
	—	40
8-inch howitzers		12
13-inch mortars	2	
10 „ „	12	
8 „ „	44	
	—	58
Total		110

Ammunition.—1000 rounds shot or 500 common shell to each piece, besides shrapnel and grape.

To supply the above, the magazines at Cawnpore, Agra, Delhi, and Kurnal, were almost drained of everything; and a reserve was formed at Allahabad in case of necessity. In like manner the stations of Benares, Allahabad, Cawnpore, Agra, Meerut, and Kurnal, were entirely drained of artillerymen; even Nusseerabad had to supply a company of golandáz, and every available artillery officer from Dinapore to the Sutlej, including some on the staff employ, were sent to join. Yet the number of foot artillery officers, captains and subalterns (not including staff), only amounted to 23; and though the horse artillery were sent into the batteries, there was not an entire relief for the number required to work the guns. The six great batteries required seven captains:* two were commanding batteries on the west

* Nos. III., IV., VI., VII., VIII., and IX., which, when it was increased to thirty-two mortars, had two captains.

1825 December

face, and three were on the staff, leaving only two of that rank for a relief.

On the 9th of December the troops began to march forward from Agra and Muttra. The town of Bhurtpore had been somewhat enlarged since the last siege; its ramparts had also been made wider and more durable, the earthwork being strengthened and bound together by wooden beams. To the north-west, and distant from the town a little more than a mile, was the Moti Jhil, an irregularly shaped lake secured by a *band* or dam, from which a watercourse led directly to the ditch, which could thus be filled in a few hours, and the defensive value of the fort be thereby greatly increased, as we found in 1805. But the water was still all in the lake; and therefore Major-General Reynell was, on the morning of the 10th, sent forward with a troop of horse artillery (Brigadier Stark accompanying it), two squadrons 11th Dragoons, 4th Bengal Cavalry, Skinner's Horse, two companies 14th Regiment, a regiment of native infantry, and a party of sappers and miners under Lieutenant A. Irvine, of the engineers. The force proceeded through the jungle which surrounds the town, and came upon a party of the enemy's horse. A few rounds from the horse artillery obliged them to retreat, but the town opening fire they retired. Meanwhile, Colonel R. Stevenson, quartermaster-general, with the sappers had gone straight towards the dam. It had only just been cut, and the sappers were able to close up the breach without much difficulty. A party from the Agra division had also been sent forward about the same time on a similar duty. Major Whish and a troop of horse artillery accompanied it.

See Plate No. XLV.

The position taken up by our force will be seen by a reference to the plan. The arc which it formed was

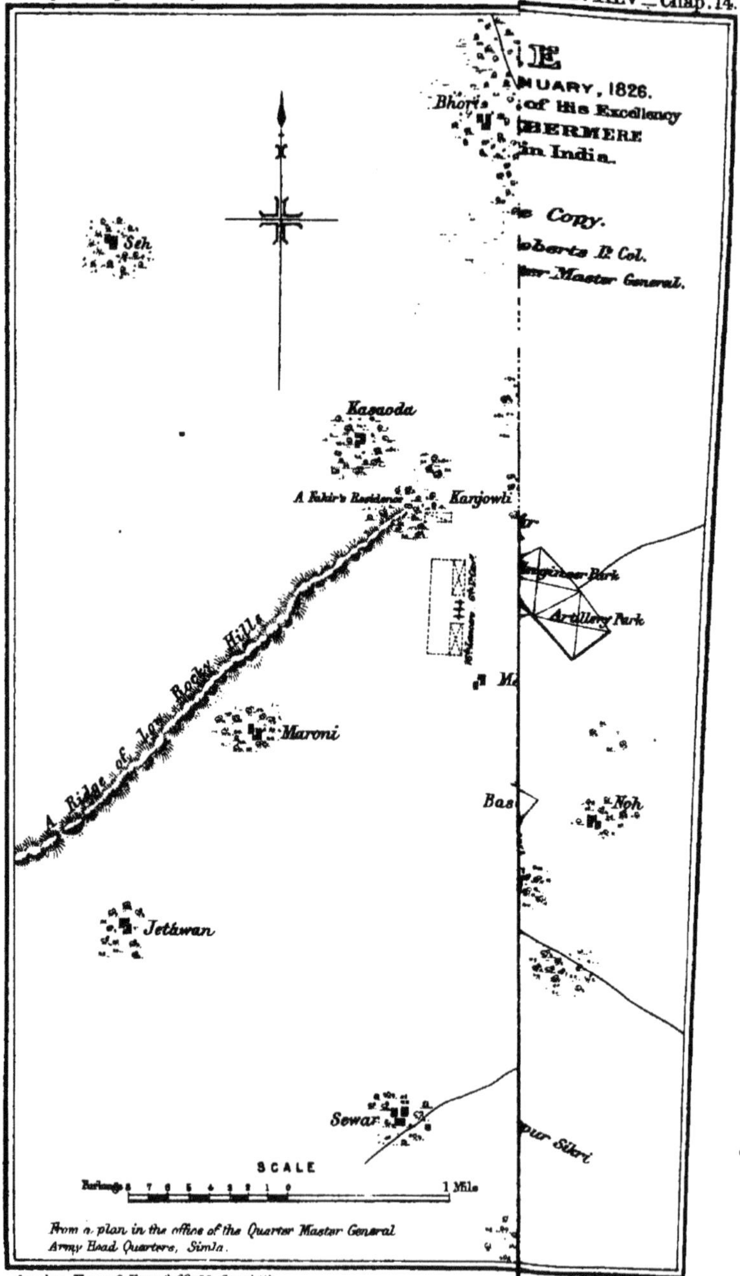

E
NUARY, 1826.
of His Excellency
BERMERE
in India.

Copy.
oberts Lt Col.
ter Master General.

Bhoris
Seh
Kasaoda
A Fakir's Residence
Kanjowli
Engineer Park
Artillery Park
A Ridge of Low Rocky Hills
Maroni
Bas
Noh
Jetawan
Sewar
pur Sikri

SCALE
Furlongs 8 7 6 5 4 3 2 1 0 1 Mile

From a plan in the office of the Quarter Master General
Army Head Quarters, Simla.

London: Henry S. King & Co. 65, Cornhill.
Edwd Weller, Litho

1825 December

necessarily so extensive that the line was in parts only imperfectly secure. Each arm of the service had its own share of the work to be performed, which was in no case slight. The infantry had, in addition to the guards and patrols for the trenches and camp, to furnish large working parties, which were employed of course by night as well as by day. The native infantry, after the duties assigned to them had been reduced by field general orders of the 23rd of December to the smallest limit, supplied as guards only, not including orderlies for regimental, divisional, and general purposes, more than 2600 native officers, non-commissioned officers, and men,* relieved weekly.

On the night of the 11th Lieutenant-Colonel Faithful was sent to occupy Mallai, a village south of the town, and in a commanding position. From this post a line of *abatis* was constructed to the right and left through the jungle, to secure and cover a road of communication.

The battering-train arrived in different detachments on the 13th, 14th, and 15th. Two 12-pounder field guns were placed at a post near the Moti Jhil, to protect the dam, and two 6-pounder guns on the right of Mallai. No time was lost in providing the necessary siege *matériel*, of which a large quantity had already been made up. Working parties were detached to assist; they were paid at the rate of four annas (sixpence) a day to each European, and a free ration of grain to each native soldier.

The point selected for attack was the north-east angle, and on Brigadier Anburey reporting his preparations sufficiently complete, possession was taken of Baldeo

* See Note D in the Appendix. Modern computation would not call this the smallest limit.

1825 December

See Plate No. XLVI.

Singh's garden and of the village of Kadamkandi as supports for the flanks of the first parallel. In front of these posts two batteries were constructed; No. I., for eight 18-pounders, was opposite to Kadamkandi, and No. II., for four 10-inch and twelve 8-inch mortars, opposite the garden.

December 22nd.—Regular foraging parties from the cavalry and infantry were sent out daily from this date. A party of 120 dismounted horse artillerymen, with their officers, were sent to the park to assist there. To facilitate communication between the park and trenches two roads were made for carts, etc., to go and return by.

December 23rd.—The army changed ground this morning, and closed up to the trenches, to cover the works more effectually. Four officers and 400 men from the artillery were told off to carry and lay down the platforms in the evening. The garden and village were occupied in force, and a party of cavalry and infantry posted on the left, to keep open the communication between Kadamkandi and Colonel Faithful's post at Mallai. During the night the guns and mortars were taken down to the batteries. The detail of officers for them was ordered:—One field officer and one adjutant, one captain, seven subalterns of horse and the same number of foot artillery.

December 24th.—Ten mortars and eight guns were in the batteries, and the fire opened at daybreak, the former directed against the town and citadel, the latter on the defences.

No. I. battery, manned by the 2nd Company, 1st Battalion, kept up a warm fire from the 18-pounders; and the practice of Lieutenant A. Abbott, with shrapnel, drew forth the admiration of Lord Combermere. The fire from the town was occasionally well directed. The

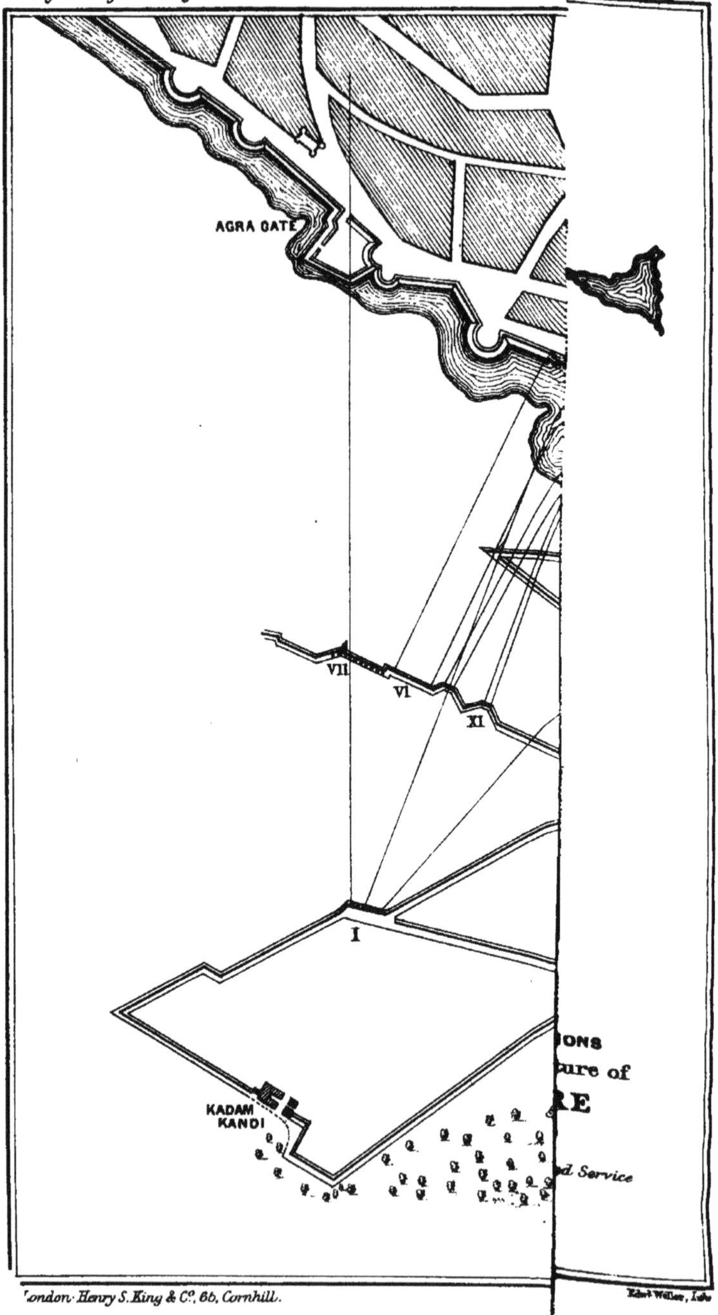

AGRA GATE

VII

VI

XI

I

KADAM
KANDI

IONS
ture of
RE

d Service

London: Henry S. King & C°, 65, Cornhill.
Edwd Weller, Lith

enemy had the range of the garden, and kept its garrison on the alert. Two 6-pounders were sent there, and two 12-pounders to Kadamkandi. The armament of the two batteries was completed in the evening by four 10 and two 8-inch mortars.

During the night the engineers commenced upon a third battery in advance (No. III.), but the work was much interrupted by the enemy's fire. Captain R. Smith, who was directing the works, received a severe contusion from a jingal shot. The battery was not ready in the morning in consequence, and the guns (five 24 and five 18-pounders) were left in the garden.

December 25th.—The trench of communication with the advanced battery was improved to-day, while the batteries kept down the enemy's fire. Captain Curphey's field battery was sent to Kadamkandi, and two 12-pounder and two 6-pounder guns from the horse artillery to the garden. A double ration of rum was served out to the British soldiers in honour of the day, and 2 lbs. of sweetmeats were served out to the sepoys at work in the trenches,* and ordered to be continued as a daily issue. The relief for the working parties paraded daily at 4 p.m. The casualties from the 23rd to this day were two gunners of the horse artillery wounded, one golandáz killed, and another wounded.

* It does not appear that any objection was made by the sepoys to working in the trenches, though it did militate against their ideas of caste. Probably the mutinous feeling exhibited by the 15th Native Infantry, terminating in the "ominous squall" mentioned by Sir Thomas Seaton,[1] had its origin in this feeling. A sepoy of that regiment, badly wounded, had had a vein opened in the arm and temple without any successful result. His body was carried by his comrades and those of the same caste through the camp of the 6th Brigade, while they exclaimed, "See, this is the way we are cut up in hospital." There was more in these words than met the ear. It never has been in the nature of the Hindustani sepoy to give his real complaints direct expression.

[1] "Cadet to Colonel," vol. i. p. 71.

1825 December

December 26th.—No. III. battery, completed and armed this morning, opened at eight o'clock. The fire, being very efficiently delivered, soon silenced the enemy's opposing ordnance, which was withdrawn to the inner defences, whence a reduced fire was kept up. Two 8-inch howitzers were sent to the battery at Baldeo Singh's garden, whence the two 12-pounder horse artillery pieces were ordered to Kadamkandi, to replace two of Captain Curphey's which were withdrawn.

December 27th.—The enemy's fire was more active and better aimed this day. A bombardier of the 4th Company, 3rd Battalion, named Herbert, had deserted the day before, and was seen on the ramparts of the town laying their guns.* The engineers had commenced on a battery (No. IV.) for eight 24-pounders and four 18-pounders, about 350 yards from the north face of the north-east angle of the town. From the right of Baldeo Singh's garden an old dyke (*a a a*) extended, which was converted into a trench, and continued up along the rear of the new battery. The fire of the enemy, however, continued during the night, and greatly impeded the work. In the afternoon the guns were withdrawn from No. I., which was to be converted into a mortar battery. All the field-pieces also were withdrawn from the garden and from Kadamkandi, except two 12 and two 6-pounders left at the former place.

Most of the enemy's horse was left outside the town; it was supposed from an apprehension that they would consume too much grain. It was reported that they would endeavour to break through the line of cavalry regiments which, under Brigadier-General Sleigh, extended from the lake to the village of Mallai. Major Whish, with the two troops of the 1st Cavalry Brigade

* See Note E in the appendix to this chapter.

under Captains Roberts and Hyde,* was under General Sleigh's orders. He was sent out each morning with an escort of cavalry to clear the jungle. On the 25th the whole brigade moved down towards the Anah gate, and inflicted severe loss on the enemy. Several skirmishes took place on this side of the town; but notwithstanding the precaution taken, a party of them made good their escape on the night of the 26th.

1825 December

December 28th.—No. IV. battery was armed and opened fire at 3 p.m.

No. I. was armed with ten 10-inch mortars, and the whole were ordered as follows:—

No. I. to fire into the town south of the citadel.
No. II. to fire into the citadel and to the west of it.
No. III. to enfilade the east face of the citadel.
No. IV. to enfilade the west face.

The whole of the artillery officers, except those under Major Whish, were placed on one roster for duty.

December 29th.—A battery of two 8-inch howitzers was formed near the village of Akad, and, under the command of Major Whish, continued to fire during the siege. It was next day increased to six pieces, and was manned by the horse artillery of the first brigade.

The other batteries kept up a heavy fire throughout the day.

December 30th.—The ground on the right of No. IV. not being good, this battery was extended to take in four more guns. During the previous night a battery (No. V.) for two guns was constructed to enfilade the ditch opposite the breach to be made by No. IV. It was

* I give these names chiefly on the grounds that they belonged to the 1st Cavalry Brigade, but have not been able to ascertain them with absolute certainty. Colonel D. Ewart served, he tells me, for the most of the time in No. IX. battery; but his troop, Colonel Duncan informs me, was with Whish. All the subalterns that could possibly be spared were sent into the trenches.

1825 December

armed and opened this day under command of Lieutenant A. Abbott.

December 31*st.*—The engineers constructed during the night a new battery (No. VI.), which was armed with five 24 and five 18-pounders on the left of the attack. The last-named guns were drawn from No. III., their places being supplied by six 8-inch mortars from No. II.; the four additional guns for No. IV. were sent there in the afternoon. During the night No. VI. was extended to the left to form a battery (No. VII.) for ten 10-inch mortars. The firing on this and the previous day was slack, as the large expenditure on the 29th had produced considerable effect.

1826 January

January 1*st.*—Major Whish's battery opened this day. Working parties were now very busy and were frequently relieved. The mortars for No. VII. were sent there from No. I. Another battery for ten 8-inch mortars (No. VIII.) was made to the right of No. IV. It was completed during the night. Lieutenant Tindall, of the engineers, was killed this evening, while marking out a site for one of the battery magazines.

January 2*nd.*—There was very little firing this day. The centre mortar battery (No. IX.) was made in the trench connecting the right and left attacks. Its armament, two 13-inch and six 8-inch mortars, was afterwards increased by twenty-four more of the latter. Two 8-inch howitzers (No. X.) were placed between Nos. IV. and VIII., and two more (No. XI.) to the right of No. VI.

The front of attack being now so extensive, it was necessary to divide the batteries into wings, each under a field officer. A depôt of ammunition was formed in the old mortar battery at Baldeo Singh's garden, and placed under a conductor of ordnance. Stores were conveyed to this depôt in carts, and thence to the batteries by working parties :— 1826 January

January 3rd.—At seven o'clock this morning all the batteries opened fire.

No. III.—Destroying defences and shelling the citadel.
No. IV.—Breaching.
No. V.—Enfilading the ditch and face of the town.
No. VI.—Breaching and destroying the defences left of the long-necked bastion (*b*).
No. VII.—Shelling the town south of the citadel. *
No. VIII.—Shelling the town between the citadel and Jangina gate.
No. IX.—Shelling the citadel and the breaches.
No. X.—Assisting No. VIII.
No. XI.—Enfilading the long-necked bastion (*b*).
Major Whish's battery.—Shelling the citadel.

January 4th.—The breaching batteries were very active this day. A sap was commenced to run along the counterscarp (*c c c*), and from thence a mine leading under the north-east bastion (*d*).

January 5th.—The trenches were extended to the right as far as a small piece of water opposite the Jangina gate, where a battery for two 18-pounders (No. XII.) was marked out. Another battery (No. XIII.) in the same trench, and about fifty yards to the right of the mortars, was also begun upon.

A difficulty was experienced in conveying sufficient ammunition to the batteries to maintain the heavy fire now kept up, and working parties from the native infantry of 150 men, relieved thrice during the day, were employed for this purpose. Officers commanding batteries were ordered to make timely application to the

* Lord Combermere had, on December the 21st, sent a letter to Durjan Sál, offering a safe conduct for the women and children, that they might escape the consequences of a bombardment, but received only an evasive reply. To a second offer no answer was returned.—Commander-in-chief's despatch to Governor-General, dated December 23rd, 1825.

1826 January

field officers for their requirements; and the latter, if they had not the means themselves of transporting the ammunition needed, were ordered to obtain the assistance of the fatigue details.

Orders were issued detailing the storming parties, as it was expected that the mine would complete the opening for an assault. Heavy firing from the town to-day.

January 6th.—Battery No. XII. was armed with two 18-pounders at 2 p.m. to-day, and Lieutenant Archdale Wilson placed in command of it. The expenditure from the batteries this day was very heavy. The right breach (*e*) appeared to be practicable, but the left one (*f*) looked very unpromising, as the ditch here was very deep, and the shot, burying themselves in an earthen mass strengthened with logs of wood, had not much effect. An attempt was made to improve the former by a mine, which was fired at dawn this morning; but the gallery had not been completed, under the apprehension that the work, if carried on by day, would be discovered. It was therefore a failure. A number of natives accustomed to trench work had been entertained to assist the sappers. Among these, it appears, was the man who in 1807 had directed the countermining operations at the siege of Kamonah which, it will be recollected, were so successful.[1] He was now in the service of Ahmad Baksh, the guardian of the young Rájá of Alwar. The men received pay at the rate of four rupees per foot.[2]

January 7th.—About five o'clock this morning the mine under the bastion *d* was fired; but this, from not not having been carried far enough home, and being charged with too small a quantity of powder, was not as

[1] Letter from Dr. Sandham.—*East India United Service Journal*, vol. v. p. 334.

[2] Letter from Lieutenant John Fisher, Sirmur Battalion.—*East India United Service Journal*, vol. v. p. 337.

successful as it should have been. A jemadár of the sappers, Barjur Singh, finding the ignition of the train stopped by some accident, ran up with a light and fired it, getting severely burned by the explosion. For this act of bravery he was promoted to the rank of subadár in general orders.

January 8th.—Durjan Sál made overtures this day, but they did not imply unconditional surrender, which was demanded, and were not accepted.

The orders for the storm which should have taken place to-day were countermanded, Brigadier Anburey having reported the breaches not sufficiently practicable. Some alterations were made in the arrangements, and Captain H. J. Wood received orders to be ready with two 12-pounders to blow open the Agra gate for the entrance of the column under General Nicolls, in case previous efforts should fail.

To ascertain the condition of the right breach, Lieutenant-Colonel Stark and his adjutant, Lieutenant H. Garbett,* ascended it last night nearly to the top, and found it practicable. The left breach still appeared too difficult for any attempt.

January 9th.—About two o'clock this morning a serious explosion, happily attended with but little loss of life, occurred in the depôt at Baldeo Singh's garden, from a shot having struck a loaded waggon. A large quantity of fixed ammunition and engineers' stores were destroyed. Captain G. Brooke, deputy commissary of ordnance, was sleeping under a bank close by, and had a narrow escape. Brigadier Childers with Lieutenant Huthwaite, accompanied by an escort, went out to select a spot to the south-east, at which a sand-bag battery for field guns was to be constructed, with the object of dis-

* "Journal of the Siege," but the name is not mentioned.

1826 January

tracting the attention of the garrison during the assault. It was placed at the edge of the jungle, about 400 yards from the south-east corner of the town.

The counterscarp opposite the right breach (*e*) was blown in this afternoon, and a good road into the ditch was thus opened.

The 1st Bengal European Regiment joined the army this day, and was encamped in front and to the left of the line, between the 31st Native Infantry and the 11th Dragoons; the volunteers for the storm, which had been called for from the cavalry, were therefore directed to rejoin their respective regiments.

January 10*th.*—The fire from our batteries, at first heavy, slackened during the day. The enemy were countermining, but one of their galleries was detected and blown in.

January 11*th.*—Two 12-pounders were added to Lieutenant Wilson's battery, No. XII., and the two howitzers in No. XI. were changed for guns. Some of the enemy's having occupied the ditch, Lieutenant J. Fisher with a party of his Gurkhas was sent in to clear it, and Captain Bell was afterwards ordered to drop shells into it from No. VIII. with reduced charges. Here small mortars would have been found useful, but there were none. Fortunately no casualties occurred among our men from splinters at so short a distance. The enemy were engaged in covering the Jangina gate with an outwork.

January 12*th.*—Another party of the Sirmur battalion, under Lieutenant J. Fisher, accompanied Captain Taylor and Lieutenant Irvine, of the engineers, to clear out the enemy from the ditch on the northern face. It was discovered that a parapet of cotton bags (*g*) had been placed across the ditch, as a traverse to stop the shot from

No. V. Beyond this, a gallery was found running under the rampart, which, it was supposed, communicated with a mine under the breach. A number of holes in the counterscarp were found filled with bodies. It was therefore determined to destroy this gallery.

1826 January

The engineers began to-day to drive a gallery from the ditch under the long-necked bastion *f*, and to sink a shaft from the sap *c c* for a gallery under the bastion and cavalier *d*. Strict orders were issued to prevent any person but those engaged in the works from entering into the trench to prevent it becoming known.

At nine o'clock in the evening, Captain Taylor (engineers) went with a party of the 14th Foot to destroy the gallery referred to above. Captain W. S. Bertrand, with Ensign W. L. O'Halloran, commanded the party. Captain Taylor, dressed in a drab greatcoat and a lascar's cap,[1] on coming near the traverse, went forward with one or two men, and getting over it was attacked by the enemy, of whom there were a number in the gallery. The rest of the party came up; but unfortunately, in the struggle which took place at the mouth of the place, Captain Taylor, not being recognized by reason of the darkness, and probably also of his dress, received some severe bayonet wounds. The attempt to blow up the place therefore failed.

January 13*th*.—The mining was continued. Batteries slackened fire during the day. Much disappointment was experienced in camp at the delay of the storming, which report daily postponed for another twenty-four hours or so. Men's minds, wound up into a state of anxious preparation for the final effort, were becoming restless under it. This will account for the number of

[1] Letter from Mr. M.—*East India United Service Journal*, vol. vi. p. 441.

1826 January

officers who were, it is stated,[1] in trouble for various misdemeanours at this time. None, however, appear to have been serious cases. Hope deferred is one of the experiences which it behoves a soldier to inure himself to, as for any of the other contingencies of military service.

January 14th.—The mine under the long-necked bastion was fired this morning, but prematurely, as it had not gone far under it;[2] little effect therefore was produced, though two guns on the rampart were brought down by the artillery fire. Lieutenant Irvine (engineers), with a party of sappers* and a few men of the 14th, at 9 a.m. succeeded in blowing in the gallery *h* to a certain extent; and Captain Farrington, commanding No. IV. battery, pouring in a steady fire immediately afterwards, opened a small but good breach.

January 15th.—Owing to the quantity of *kankar* † in the soil, the gallery under the counterscarp leading to the left breach was abandoned, and the enemy took the opportunity of the temporary suspension of operations to scarp the foot of the breach. The battering at this point was again resumed. Lieutenant H. de Bude (engineers) was very severely wounded during the night by a matchlock ball.

January 16th.—No. VI. firing salvos at the left trench. No. IV. (Captain Farrington) destroyed another gun which bore upon the right breach.

A supply of ammunition was this day received from the Agra magazine, which now had no more in store.

* Conductor Richardson headed the sappers. The party, small as it was, dislodged the enemy and held the mine till it was loaded with 400 pounds of powder.

† A rapidly forming nodular limestone found near the surface.

[1] Letters, *East India United Service Journal*, vol. vi. p. 412.

[2] Letter from Lieutenant J. Fisher.—*East India United Service Journal*, vol. vi. p. 412.

1826 January

The great mine under the long-necked bastion was fired about four o'clock p.m. Immediately afterwards Captain Carmichael (59th Regiment), aide-de-camp to General Nicolls, with six grenadiers of his regiment and four of the Gurkhas, under a heavy fire from the trenches, ran up to the top of the breach, and after looking well at the inside and throwing some hand grenades, returned without further loss than one of the grenadiers, who was killed just as they got back. Captain Davidson, of the engineers, also accompanied the party.

January 17*th*.—The second great mine, that under the north-east angle, was now in the centre of the bastion and under the cavalier. It was this day charged with 10,000 lbs. of powder. Its explosion the following morning was to be the signal for assault.

For this two main columns and three smaller columns of attack had been ordered—one, under Major-General Reynell, for the right; the other, under Major-General Nicolls, for the left breach, *i.e.* the bastion. A column commanded by Lieutenant-Colonel J. Delamain, 58th N.I., was directed to the breach near the Jangina gate made by Lieutenant Wilson's guns; one led by Lieutenant-Colonel T. Wilson, 33rd N.I., to escalade the re-entering angle of the long-necked bastion (*k*); while Brigadier-General J. W. Adams was to effect an entrance at the Agra gate. The cavalry were disposed by Brigadier-General Sleigh along the whole of the west side as far as Mallai. The two major-generals each had a troop of horse artillery placed under their orders, and a spiking party of a sergeant and sixteen rank and file from the artillery accompanied each column. The whole disposable force in camp was employed either for the storm, for the camp and trench picquets and guards, or to follow up the successful result.

1826 January

January 18*th*.—From the earliest possible hour this morning, a heavy and uninterrupted fire was kept up in all the batteries, almost drowning the concussion caused by the springing of the mine under the north-east angle and cavalier. For a few moments after this the fire redoubled its intensity, as the columns cleared the trenches and rushed forward to the assault, and then ceased. The work of the siege artillery before Bhurtpore was finished.

NORTH-EAST ANGLE BASTION.
(The Breach at which Major-General Reynell's Column entered Bhurtpore.)

The explosion of the mine made a splendid breach, but General Reynell's column was so close that Brigadier-General McCombe and Brigadier R. Paton, Lieutenants Irvine (Engineers) and R. Daly (N.I.), were severely wounded. Lieutenant-Colonel S. Martin, 23rd N.I., the next senior officer, went to the front, but was wounded at the top of the breach. A mine sprung by the enemy here blew up a section of the 14th Regiment, but it did not stop the rest. Owing to the firmer nature of the ground in the crater of the mine, the whole of the column went up by the breach in the bastion, in preference to that made by the batteries in the adjoining curtain. On reaching the top the column divided according to order, sweeping the ramparts right and left.

The column under Lieutenant-Colonel Delamain

LONG-NECKED BASTION.

(The Breach at which Major-General Nicolls' Column entered Bhurtpore.)

mounted the breach near the Jangina gate, in spite of a determined opposition.

The other main column, under Major-General Nicolls, mounted the breach in the long-necked bastion. The enemy at the summit had brought their guns to bear upon it. Brigadier-General Edwards fell desperately wounded, and a stubborn resistance had to be overcome before the enemy were compelled to give back. Their gunners were shot or bayoneted at their posts.

1826 January

Lieutenant-Colonel Wilson, as had been intended, attacked the re-entering angle of the long-necked bastion; Lieutenant Anderson, with his pioneers, successfully planted the ladders. This column suffered severely, but was in its place upon the top of the ramparts in time to join their comrades. Sweeping round the walls, they opened the Agra gate for the party under Brigadier-General Adams, and continued their course until, at the Fateh Burj, they met the head of General Reynell's column.

Till 12 o'clock the fighting continued in the town; the citadel held out a little longer, but at 3 p.m. a white flag on the walls terminated all resistance, and shortly after Durjan Sál brought in by a detachment of cavalry completed the day's success.

Brigadier-General Sleigh had subdivided the two troops of horse artillery along his whole line: two guns in front of Jhilai, two on the road from Anah to Sewar, two guns and four howitzers to the left of Goálpáráh. Besides these, a couple of squadrons under Captain J. Jenkins, 11th Dragoons, were placed in support of Lieutenant Huthwaite's battery to the front of "Colonel Faithful's post," which drew down upon itself a heavy fire from the town.

Bhurtpore having been secured, Brigadier-General McCombe was left in charge, and the army moved against the other fortified places belonging to the *ráj*, Biáná, Wer, Kumbher, Deeg, and Kámá, which were all given up without resistance being attempted.

The casualties among the artillery at this siege were very trifling, only amounting in the ranks to one horse artilleryman killed and three wounded; nine foot artillerymen killed and eighteen wounded. The returns, which are very defective, do not give either grade or nationality. Besides Captain Brooke, who was mentioned above as slightly injured by the explosion, Lieutenants Macgregor and Maclean were both wounded, the latter severely; but he is not included in any list.

A large booty was secured by the prize agents, of whom Major W. Battine was one; also 132 pieces of ordnance* of all kinds, and 300 wall-pieces. Two large bronze guns were selected—one, called "Matsad Ali," for presentation to his Majesty the King; the other was given to the head-quarters of the artillery, and is now with other trophies of the old regiment at Woolwich. A sketch of it is given in page 216, taken while it was at Meerut. The carriage was made for it in Fort William.

In the official despatches, Brigadier-General J. Sleigh records the names of Lieutenant-Colonel Stark and Major Whish; the commander-in-chief, the names of the commandant of the regiment, of Brigadiers Hetzler and Clement Brown, with honourable mention. The last-named officer was subsequently gazetted, among other officers, a Companion of the Bath. He, as well as Lieutenant-Colonel Stark, had been at the former siege of Bhurtpore, witnesses of our failure and our success. Captain James Tennant, who, as adjutant-general of artillery, had the management of all details connected with the artillery generally, was thanked by the commandant in regimental orders[1] for the assistance he had rendered to him. The methodical habits and mathematical talent of this officer rendered labour easy to him which would have been difficult to others.

With the siege of Bhurtpore ended for a time the field services of the Bengal Artillery. For twelve years

* Besides a large quantity taken from the other forts. Lieutenant Garrett was detailed (G. O. February 1st) to collect and take the whole to Agra.

[1] 21st January, 1826.

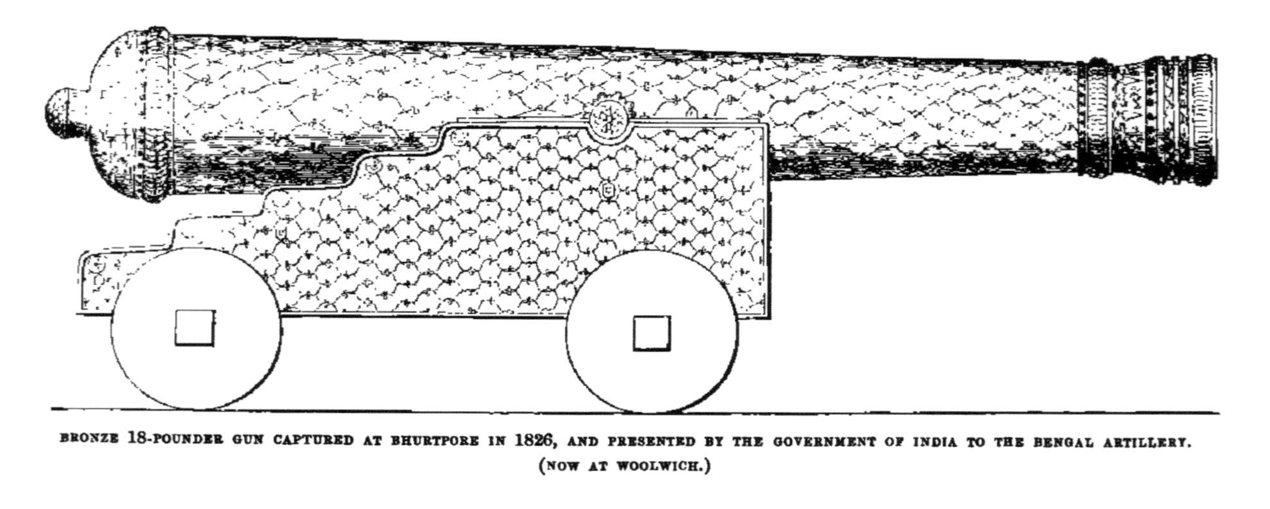

BRONZE 18-POUNDER GUN CAPTURED AT BHURTPORE IN 1826, AND PRESENTED BY THE GOVERNMENT OF INDIA TO THE BENGAL ARTILLERY.
(NOW AT WOOLWICH.)

after, its guns were not unlimbered against an enemy. The country meantime endured a stormy peace; for while the sword was sheathed, the officers of its army, instead of sitting down to read over again the lessons they had begun to learn in the Nipál and Pindári campaigns, were raving about the half-batta question, until the greater injustice of the Afghán war came to set men's minds right, by turning their thoughts into another channel.

1826 January

AUTHORITIES CONSULTED FOR THIS CHAPTER.

1. Journal of the Artillery Operations before Bhurtpore. *East Indian United Service Journal*, vol. ii.

2. Letters from officers written during the siege. *East Indian United Service Journal*, vols. v. and vi.

3. Narrative of the Siege and Capture. By J. N. Creighton, Esq., Captain 11th Regiment Light Dragoons. 1 vol. 4to. London, 1830.

4. From Cadet to Colonel. By Major-General Sir Thomas Seaton, K.C.B. 1 vol. 8vo. London, 1866.

5. Memoir of Lieutenant-Colonel James Skinner, C.B. By J. B. Fraser. 2 vols. 8vo. London, 1851.

6. Letters from Generals Sir G. Brooke, Sir J. Alexander, and J. Abbott; Colonels D. Ewart and F. K. Duncan.

7. Copies of Muster-Rolls.

APPENDIX.

Note A.—Numerical strength of the force employed at Bhurtpore in 1825-1826.

Note B.—List of artillery officers who served at the second siege of Bhurtpore.

Note C.—Numerical strength of the artillery employed.

Note D.—Detail of guards furnished by the native infantry portion of the force before Bhurtpore.

Note E.—The deserters.

Note A.

Numerical strength of the force employed at the siege of Bhurtpore in 1825-1826. From Captain Creighton's "Narrative and Journal of the Siege."

	Officers.	European N.-C. O., rank and file.	Native commissioned, N.-C. officers, rank and file.	Total of each regiment.	Total of each arm.
Artillery	71	1179	1133	2383	2383
11th Light Dragoons ...	35	596	...	631	
16th Lancers	34	616	...	650	1281
3rd Bengal Light Cavalry	13	2	456	471	
4th " ...	13	2	620	635	
6th " ...	12	2	517	531	
8th " ...	15	2	466	483	
9th " ...	13	2	594	609	
10th " ...	13	2	425	440	3169
14th Regiment	35	660	...	695	
59th "	30	637	...	667	
Bengal European Regiment (one wing)	14	462	...	476	1838
6th Native Infantry ...	12	2	1138	1152	
11th " ...	11	2	1040	1053	
15th " ...	17	2	934	953	
18th " ...	13	2	1070	1085	
21st " ...	17	2	1105	1124	
23rd " ...	13	2	1120	1135	
31st " ...	17	2	1036	1055	
32nd " ...	12	2	994	1008	
33rd " ...	15	2	1021	1038	
35th " ...	19	2	1020	1041	
36th " ...	17	2	1172	1191	
37th " ...	17	2	1058	1077	
41st " ...	6	2	566	584	
58th " ...	13	2	1146	1161	
60th " ...	20	1	1084	1105	
63rd " ...	13	2	1036	1051	16,813
				Grand Total ...	25,484

N.B.—Skinner's Irregular Cavalry, which was about 1000 strong, is not included above.

NOTE B.

List of officers of Artillery who served at the second siege of Bhurtpore, 1825-1826.

Lieutenant-Colonel Commandant Alexander Macleod, C.B., Commandant of the Regiment—Brigadier commanding Artillery.
Captain James Tennant, Assistant Adjutant-General.
Lieutenant John Stuart Rotton, Deputy Assistant Quartermaster-General.
2nd Lieutenant F. Dashwood, Aide-de-camp.

HORSE ARTILLERY.

First Brigade.

Major William S. Whish commanding.
Lieutenant Donald H. Mackay, Adjutant.

2nd Troop.
- Captain Roderick Roberts commanding.
- Lieutenant Richard Scrope B. Morland.
- " David Ewart.
- 2nd Lieutenant William E. J. Hodgson.

Second Brigade.

Lieutenant-Colonel Henry Stark * commanding.
Lieutenant James Johnson, Adjutant—Brigade-Major Foot Artillery and Battering Train.

1st Troop.
- Captain James C. Hyde commanding.
- Lieutenant Thomas B. Bingley (with 4th Troop, 3rd Brigade).
- Lieutenant Hubert Garbett, Acting Adjutant 2nd Brigade.
- Lieutenant William Anderson (from 3rd Troop, 1st Brigade).
- 2nd Lieutenant Francis Dashwood, Aide-de-camp to Brigadier Macleod.

2nd Troop (part of).
- Lieutenant Charles R. Whinfield, Brigade-Major Horse Artillery.

3rd Troop.
- Captain George Blake commanding.
- Lieutenant Charles H. Wiggens.
- " Julius B. Backhouse.
- 2nd Lieutenant Frederick Grote.

4th Troop.
- Captain John J. Farrington commanding.
- Lieutenant John Cullen.
- 2nd Lieutenant Francis B. Boileau (from 3rd Troop, 1st Brigade).

* Had served at the first siege of Bhurtpore.

Third Brigade.

Lieutenant-Colonel Commandant Clements Brown,* Brigadier commanding Horse Artillery.

Lieutenant Gervaise Pennington, jun., Adjutant.

1st Troop.
- Captain Henry J. Wood commanding.
- Lieutenant William R. Maidman.
- " James W. Wakefield.
- 2nd Lieutenant William S. Pillans.

2nd Troop.
- Captain Jonathan Scott commanding 3rd Brigade.
- " William Bell (from 3rd Troop) commanding.
- Lieutenant Charles McMorine.
- " James Alexander.

4th Troop.
- Captain Gabriel Napier C. Campbell commanding.
- Lieutenant Thomas Nicholl.
- " Thomas B. Bingley (from 1st Troop, 2nd Brigade).
- Lieutenant George Maclean (from 4th Troop, 1st Brigade).

FOOT ARTILLERY.

First Battalion.

Lieutenant-Colonel John A. Biggs commanding.

Lieutenant Robert G. Macgregor, Adjutant.

2nd Company.
- Lieutenant Augustus Abbott commanding.
- " Proby T. Cantley (joined from Canal Department).
- 2nd Lieutenant James Abbott.

3rd Company.
- Captain George H. Woodroofe commanding.
- Lieutenant John R. Revell.
- " George Ellis (doing duty with 6th Batt.).

4th Company.
- Captain George Brooke, Deputy Commissary of Ordnance.
- Lieutenant John Edwards commanding.

Third Battalion.

Lieutenant-Colonel Commandant Robert Hetzler, C.B., Brigadier commanding Foot Artillery and Battering Train.

Lieutenant Thomas Sanders, Adjutant.

1st Company (Field Battery)
- Captain William Curphey commanding.
- Lieutenant Henry P. Hughes.
- " Peter A. Torckler.
- 2nd Lieutenant Francis K. Duncan.

2nd Company (Field Battery)
- Captain Peter L. Pew commanding.
- Lieutenant Edward Huthwaite (doing duty with 6th Battalion).
- 2nd Lieutenant Thomas E. Sage.

---

* Had served at the first siege of Bhurtpore.

4th Company. { Captain Isaac Pereira, Commissary of Ordnance.
Lieutenant Edward S. A. W. W. Wade commanding.
2nd Lieutenant Elliot D'Arcy Todd.

Fourth Battalion.

Major William Battine commanding.
Lieutenant Richard Horsford, Adjutant.

2nd Company. { Lieutenant Archdale Wilson commanding.
2nd Lieutenant Francis R. Bazely.

3rd Company. { Captain William Oliphant, Assistant Secretary to the Military Board).
Lieutenant Rowland C. Dickson commanding.
2nd Lieutenant George J. Cookson.

Sixth Battalion.—Golandás.

Lieutenant-Colonel Charles Parker commanding.
Lieutenant Henry Clerk, Adjutant.
Lieutenant John S. Rotton, Interpreter and Quartermaster.

3rd Company. Lieutenant Henry Clerk, Adjutant in charge.
4th Company.
5th Company.
13th Company.

17th Company. { Lieutenant Edward Huthwaite, from 2nd Company, 3rd Battalion, commanding.
Lieutenant George Ellis, from 3rd Company, 1st Battalion.

OFFICERS DOING DUTY, BUT WHOSE PLACES HAVE NOT BEEN ASCERTAINED.

Lieutenant William T. Garrett, from 1st Company, 4th Battalion.

Lieutenant F. S. Sotheby (1st Company, 2nd Battalion), from Nizám's contingent.

NOTE C.

Numerical strength of the Artillery employed at the second siege of Bhurtpore. From the "Journal of the Siege" published in the *East Indian United Service Journal.*

Horse Artillery.	European non-commissioned officers, rank and file	561	
	Native ditto	212	
	Lascars	209	
			982
Foot Artillery.	European non-commissioned officers, rank and file	618	
	Native ditto	435	
	Lascars	277	
			1330
	Total		2312

Note D.

Detail of guards furnished by the native infantry portion of the force before Bhurtpore, as laid down in field general orders of the 23rd December, 1825.

	Native officers.	Havildars.	Naicks.	Drummers.	Sepoys.
Regimental Guards.					
Quarter or Standard Guard	1	2	2	2	20
Rear Guard	...	1	1	...	12
Ammunition Guard	...	...	1	...	4
Commanding Officer	...	...	1	...	4
Hospital	...	...	1	...	4
Bazar	...	...	1	...	4
Mess	...	...	1	...	4
Total	1	3	8	2	52
Division Guards.					
Two General Officers commanding divisions ...	2	2	4	...	40
Nine Brigadier-Generals and Brigadiers ...	...	9	9	...	108
Four Assistant Adjutant and Quartermaster Generals	...	...	4	...	16
Commissariat Depôt	...	1	1	...	12
Two regiments European Infantry	...	2	4	...	32
Total	2	14	22	...	208
General Guards.					
1st Division.					
The Commander-in-Chief	4	12	12	4	150
Sir Charles Metcalfe	2	6	6	2	100
General duties of Horse Artillery	...	1	2	...	16
11th Dragoons	1	1	2	...	24
16th Lancers	1	1	2	...	24
Provost Marshal	1	2	2	...	20
Head-quarters Bazar	...	1	1	...	12
Military Secretary to the Commander-in-Chief	...	...	1	...	4
Quartermaster-General H.M.'s Forces ...	...	...	1	...	4
Adjutant-General H.M.'s Forces	...	...	1	...	4
Judge Advocate-General	...	...	1	...	4
Field Paymaster	...	...	1	...	4
Superintending Surgeon	...	...	1	...	4
Quartermaster-General, Company's Service ...	...	...	1	...	4
" for General Purposes	1	2	2	...	24
Hospital Guard, including Medical Store ...	...	1	2	...	16
Total	10	27	38	6	414

NOTE D (*continued*).

	Native officers.	Havildars.	Naicks.	Drummers.	Sepoys.
GENERAL GUARDS (*continued*).					
2nd Division.					
Artillery Park Guard	6	18	18	6	300
Engineer's „	2	6	6	2	100
Commissary-General's Office and Treasury ...	...	1	1	...	12
Rum Depôt	...	...	1	...	8
Total	8	25	26	8	420
ABSTRACT OF THE FOREGOING.					
Regimental Guards: Six Regiments of Native Cavalry, and sixteen of Native Infantry	22	66	176	44	1144
Division Guards	2	14	22	...	208
General Guards, 1st Division	10	27	88	6	414
„ 2nd Division	8	25	26	8	420
Total	24	132	262	58	2186

NOTE E.

Bombardier Herbert had been in the Royal Artillery, had been present at Waterloo, and had borne a fair character. Unfortunately, he was not alone in his infamy; two gunners, Henessy and O'Brien, also deserted during the siege. Slaves to drink, they knew no other master, but it is not known what could have induced the bombardier thus to sacrifice himself. He was hanged on the bastion of the north-east angle; the other two were each discharged with ignominy from the service, and sentenced to transportation for fourteen years.

APPENDIX.

COLONEL THOMAS DEANE PEARSE.

COLONEL PEARSE was descended from a good family resident in Berkshire.* His father was a grandson by the mother's side of Chancellor Hyde, whose daughter, Anne Hyde, was married to James II. At the age of fifteen, he was admitted a cadet into the Royal Academy at Woolwich, and on the 8th June, 1757, obtained his commission as a lieutenant-fire-worker in the Royal Artillery. He was employed on service, as he says in a letter to a friend of his, Lionell Darell, Esq., at St. Malo, Cherbourg, and St. Coss in 1758, Martinico and Guadaloupe in 1759, Belleisle in 1761, and Havannah in 1762. His marked merit and abilities gained him the friendship, among others, of Generals Desaguiliers and Pattison, R.A., with both of whom he long kept up a correspondence.

In 1768 he was selected by Lieut.-Colonel James Pattison, then Lieut.-Governor of the Royal Academy, as fit for the post of commandant of the Bengal Artillery, though then only a lieutenant, in consequence of the application of the Court of Directors; and he was accordingly promoted to the rank of major in the Company's service, and came out to India. The story of his supersession, first by Major Fleming Martin, secondly by Major N. Kindersley, both originally from the Royal Artillery, the latter junior to him as a lieutenant, will be found in the volume of this work which treats of the

* The name was pronounced *Perse*, and is of Norman origin.

organization of the corps. On the death of Major Kindersley, on the 28th of October, 1769, he was promoted to the rank of lieut.-colonel, and the command of the regiment.

It would be impossible in the limits of a short sketch to record the many services rendered by Colonel Pearse to the corps he commanded for twenty-one years, or do more than refer to the obstacles he encountered. They will be seen in their places in this work.

On coming out to India he formed a friendship with Warren Hastings which lasted throughout his life; but the connection made him many enemies. He acted as his second in the duel which Mr. Hastings fought with Mr. Philip Francis, on Thursday, the 17th of August, 1780.

In 1781, Colonel Pearse was placed in command of a detachment of five N.I. regiments, which, with a newly raised company of native artillery, marched from Midnapore to join Sir Eyre Coote in the Carnatic. In this command he met with many difficulties. His determination to support the authority of majors commanding regiments, not an article of belief among their juniors, who were disposed to assume a complete independence in the management and internal economy of their companies, led to much insubordination, which was only kept down by his temper and firmness. The following extract of a letter to Brigadier-General Stibbert, Commander-in-Chief in Bengal in 1781, shows how truly Colonel Pearse was imbued with the spirit of a genuine soldier:—

"Sir,

"I am now to acquaint you that I have given leave of absence to Captain Ogilvie, to go to Masulipatam for the benefit of his health, as he has been very ill lately; and when he is there, I shall order him to return to Bengal. An officer not at the point of death, who quits his station just as he comes in sight of the scene of action, deserves no favour; and I hope, therefore, that he will never be permitted to return to the army under my command."

There have been later instances in the army to which this lesson might apply.

When Colonel Pearse joined the army of the Carnatic, Sir Eyre Coote, among whose virtues a freedom from prejudice did not exist, deprived him of the command of the force he had brought from Bengal, distributing the regiments among the other brigades; and though he was appointed soon after to another, the measure greatly mortified him, and was in more ways than one detrimental to the public interest. At the battle fought on the 27th of August, 1781, with Hydar Ali, the brigade under his command materially contributed to repair the mistakes of the day and avert defeat.

It was not till the beginning of 1785 that this detachment returned to Bengal, when their encampment was visited by the Governor-General in person, who afterwards, in G. G. O. 22nd January, tendered to them the thanks of the Government of India, for their behaviour throughout their protracted and arduous service, and granted medals of different degrees of value upon the native officers of both infantry and golandáz. An honorary sword was also conferred upon Colonel Pearse.

During this war Colonel Pearse had appointed (15th November, 1783) Lieutenant R. H. Colebrooke surveyor, and assisted him in the astronomical observations by which he laid the first foundation of a scientific survey of the country. Colonel Pearse had commenced these observations from the time of leaving Bengal. They are to be found in the first volume of the "Researches of the Asiatic Society of Bengal," p. 57. Lieutenant Colebrooke afterwards became Surveyor-General of Bengal.

After his return from Madras, Colonel Pearse employed himself, with a zeal undiminished by the many instances of unjust supersession from which he had suffered, in raising and improving the character and condition of his corps, and during his leisure hours in scientific studies, chiefly astronomical and meteorological. When the Asiatic Society of Bengal was founded in 1784, he became one of its members, and its "Researches" contain, besides the astronomical observations alluded to above, a meteorological journal (vol. i. Appendix) and a paper on two Hindu festivals and the Indian Sphinx (vol. ii. p. 333). In one of his letters to Sir Joseph Banks, president of the Royal Society, besides a meteorological journal from November, 1773, to June, 1787, he sent certain models which if still extant, would possess considerable interest. One, constructed in brass, was to show a method of working two pistons by a winch, thus communicating to wheels the motion of the beam of a steam engine. This was an original invention first made in 1773, and applied by Colonel Pearse to some fire engines which, in the following year, the Board of Ordnance requested him to have constructed. The priority or originality of the invention was called in question by Mr. Smeaton, a member of the Royal Society, but in his letter to Sir J. Banks, Colonel Pearse established his rights to the credit he claimed. Another model was of glass, a marine constructed from a land barometer, of which he gives a description; but the ingenious method of filling the tubes with mercury would not now convey, as it did then, any novel information.

The arrival in India in 1786 of Earl Cornwallis, who at once appreciated his character, was a source of much pleasure, as well as a relief, to Colonel Pearse; and though he had given up expecting what his ambition had long sought after, still, as he expressed it in a letter to Mr. Hastings, he could breathe again. On the 7th of November, about two months afterwards, the Governor-General did him the honour of paying him a visit at his "country house;" and shortly after, at his request, he laid before him a plan long before matured for saving the State the great expense of keeping up camp equipage, by the establishment of what has long been known as "tentage." This measure was subsequently introduced; and though it does not exist now as a contract for the army at large, it still forms one of the items of an officer's pay.

Colonel Pearse's health had for some time been declining, and he suffered occasionally from severe fits of illness; but he never had had the means of acquiring a sufficient fortune to enable him to retire as he had hoped to do. The large allowances given to Colonels Goddard and Morgan while holding, like him, independent commands had been withheld. And though, in his private correspondence, he complains bitterly of the way in which he had been neglected and debarred from the rank justly his due, which would have increased his private means, we do not find one sentence unbecoming a soldier.* But the current of public opinion at home was running too strongly against Hastings and those who supported him in India, and Pearse was not the only one who suffered.

On the 9th of March, 1789, a fire broke out in Fort William, by which a large number of gun-carriages, stands of arms, and a quantity of camp equipage were destroyed. Colonel Pearse was on the spot, and, it appears, was then undergoing a course of mercury (at that time considered a valuable cure for many disorders), and the exposure to the night air and the exertions he underwent may have contributed to accelerate the form of disease. But on the 15th of June following, he died at the comparatively early age of 47 years, and left a name which will always be loved and respected, as long as the regiment of which he was justly called "the father" shall be remembered.

His commissions were dated as follows:—fireworker (R.A.), 8th June, 1757; second lieutenant (R.A.), 24th October, 1761; lieutenant (R.A.), 3rd February, 1766; major, 2nd September, 1768; lieut.-colonel, 30th October, 1769; colonel, 12th June, 1779.

MAJOR-GENERAL PATRICK DUFF.

This officer was transferred to the Bengal Artillery from H.M.'s service, probably as an ensign, as the date of his original commission (12th June, 1762) was subsequently confirmed to him to adjust his standing. He was, like nearly every officer in the army, concerned in the Batta mutiny of 1766, and was, with another, tried by a court-martial for having originated in a quarrel, though accidentally, a fire, by which

* He says in one letter, dated 24th January, 1789: "Once I was high-minded and wanted a ribbon; now I shall endeavour to steer clear of a halter, since Mr. Macpherson [late Acting Governor-General] is made a baronet, and Sloper [Lieut-General Sir Robert Sloper, K.B.] has got what I strove to earn."

nearly half the bungalows* at Bankipur were burned down. He was ordered home, but was subsequently restored to the service without loss of rank. He was soon after in the field, and his conduct at the battle of Buxar elicited laudatory mention from Government. He was placed in command of a native battalion of artillery, raised for the Nawáb of Oudh, in 1776, which was soon after disbanded; and during the absence of Colonel Pearse in the Carnatic, on service, he commanded the Brigade of Artillery, as the senior lieut.-colonel in Bengal.

On the 6th of December, 1788, he left for England, and returned to India on the 3rd of December, 1790, and joined Lord Cornwallis, when he proceeded to take command of the army in the second Mysore campaign, and was present at all the important operations of that war. Being left in command at Bangalore, he prepared the battering-train for the siege of Seringapatam. After this he again returned home, as the Court of Directors declined to permit him to hold the command of the brigade as well as a battalion. This refusal may have originated in his rank of colonel,† as promotion in the place of Lieut.-Colonel C. R. Deare (killed at Sátiya

* These were then mere huts constructed of matting and thatch, situated close to each other; more an encampment than a permanent station. Such were the places in which that generation were wont to weather the heat in India.

† The order granting him the rank of colonel, and confirming him in his original standing in the service, ran as follows:—

"MINUTES OF COUNCIL, MILITARY DEPARTMENT.

"3rd December, 1790.

"The Honourable Court of Directors were pleased to order, in the 105th paragraph of their general letter of the 8th of April, 1789, that the rank and allowances of colonel should be constantly annexed to the station of the senior officer of the Brigade of Artillery on the Bengal establishment, and that Lieut.-Colonel P. Duff should be promoted to the rank of full colonel, agreeably to his original standing in the service. Colonel Duff having been permitted by the Honourable Court of Directors, under date the 29th April, 1790, to return to his duty at this Presidency without prejudice to his rank, and it appearing that the date of Colonel Duff's commission, on his first admission into the Company's service from his Majesty's, was prior to that of Colonel Christian

Mangalam) was delayed till the reply came from home, or from his junior officer, Lieut-Colonel G. Deare, having been intermediately appointed. Whatever the cause was, it had ceased to operate in 1797. He returned from furlough on the 13th of March in that year, and was on the 29th of the same month nominated to the command of the artillery, vice Major-General Deare. Government, however, immediately after (Minutes of Council dated 5th June) having ruled that general officers were not to be placed upon regimental duty, he vacated the post, in which, on the 1st of July, he was succeeded by Colonel Vere W. Hussey, and was then appointed to command the troops at the Presidency.

On the 5th of December, Major-General Duff embarked for Europe, and remained there till his death.

He was a man of very powerful frame of body. Among the anecdotes which used to be current of his strength was an encounter which he had with a leopard that charged him, and which he seized by the throat, getting, however, very much lacerated in the struggle. From this he is said to have gained the title of Tiger Duff, an affix which has, often with doubtful veracity, been given to other sportsmen since. On another occasion, finding a sentry asleep over the park, he took a 6-pounder (about 4½ cwt.) off its carriage, and walked off with it under his arm *dúrbín ke múáfik* (like a telescope), as an old native officer, who was his orderly at the time, afterwards described it.*

Knudsen, being under date the 12th of June, 1762, whereas that of Colonel C. Knudsen is dated the 27th of June, 1762:

"Resolved and ordered that a commission be made out, granting Colonel P. Duff the rank of colonel in the army from the 17th of April, 1786, immediately above Colonel Christian Knudsen, and below Colonel Arthur Ahmuty, and in the Artillery from the 29th of May the same year.

(Signed) "PETER MURRAY,
"Adjutant-General.

"Fort William,
"Adjutant-General's Office,
"20th December, 1790."

* There is a curious similarity between this and the following story:—

"The commander of Azov had placed a piece of ordnance on an outwork of the fortress, leaving a guard over it. The general commanding the forces in the neighbourhood came thither with General-in-

He had two brothers in the service: one, William, was killed at the siege of Kamonah; the other, John, retired—both in the rank of lieutenant-colonel.

The dates of his commissions were:—fireworker, 12th June, 1762; second-lieutenant, 2nd December, 1763; lieutenant, 28th March, 1764; captain-lieutenant, 2nd August, 1765; captain, 6th August, 1768; major, 2nd February, 1777; lieut.-colonel, 13th November, 1780; brevet-colonel, 17th April, 1786; regimental-colonel, 29th May, 1786; major-general, 20th December, 1793.

LIEUT.-COLONEL EDWARD MONTAGU

Belonged to a family well known in the naval annals of Great Britain. He was the fourth son of Admiral J. Montagu; and an elder brother was Captain James Montagu, who was in command of the frigate *Medea*, 28 guns, which formed part of the squadron under Vice-Admiral Sir E. Hughes, and assisted in the capture of the fort of Trincomalee, on the 5th of January, 1782; and who was killed when in command of the *Montagu*, 74 guns, in the action fought by Lord Howe with the French fleet, off the West India Islands, on the 1st of June, 1794.

After passing through the Academy at Woolwich, Lieutenant Montagu came out as a cadet to Bengal in 1770. In the first twelve months after his arrival in India he served in the "Select Picket,"* after which he was brought on the strength of the artillery as lieutenant-fireworker, 16th May, 1773.

Chief Soltikow, and finding the sentry asleep, had harnessed himself to the gun, and, together with his attendants, had brought it away without the sentry being disturbed by all the noise."—"Diary of an Austrian Secretary of Legation at the Court of Czar Peter the Great," vol. i. p. 151.

* For an account of the "Select Picket" see note, vol. i. chap. ii. p. 44.

He was employed with a force under Colonel Goddard,* and in attacking a fort was, while attempting to force the gate, wounded by an arrow, which entering below the eye passed through part of the face. Without hesitation, he broke the shaft off close to the barb, and continued at the head of his men till the object of the attack was gained. After some days the barb was skilfully extracted by Dr. Brinch Harwood.†

Captain Montagu served on the Coromandel coast in the war with Hydar Ali under Sir Eyre Coote, and upon all occasions distinguished himself, particularly at Cuddalore, in June, 1783, where his judicious disposition of the artillery of one of the wings of the army under his command was honourably acknowledged by the French officer who was opposed to him.

On the breaking out of the war with Tippoo Sultán in 1790, Captain Montagu was again sent to the Madras Presidency with his company, the 2nd Company, 2nd Battalion (not now existing), and while on this service was promoted to the regimental rank of major. Though the junior field officer of this arm, he was selected to command the artillery employed in the reduction of the forts of Nandidrug and Rahmándrug; and the fact, as stated by his biographer, that Lord Cornwallis had at first thought of entrusting the operations against Nandidrug to Colonel D. Smith, of the Madras Artillery, but afterwards altered his intentions in favour of one considerably his junior,‡ may perhaps account for the omission altogether from

* In the "Asiatic Annual Register" for 1799 (vol. i. p. 65), from which most of this sketch is taken, this is said to have occurred in the Rohilla country, in 1781. But Goddard was employed against the Máhrátás then. He was, however, in Rohilkhand in 1773, when an English force co-operated with the Vazir of Oudh and Háfiz Rahmat Khán in expelling the Máhrátás from that country ("East India Military Calendar," vol. ii. p. 415). It is possible that Montagu, who was commissioned as lieutenant-fireworker in May, 1773, may have served with that force in that rank. Besides, he was borne on the returns of one of the companies which sailed with Sir Eyre Coote to Madras in October, 1780. Captain Buckle's supposition that Lieutenant Montagu was quarter-master to the artillery in 1781 is most probably correct.

† Professor of anatomy in the University of Cambridge in 1800.

‡ Lord Cornwallis did not appear to have meant any reflection upon Colonel Smith by this; he evidently merely selected the best man, and wished to have him unfettered in his judgment by not sending another with him who was his senior. See note E, vol. i. chap. iv.

the despatches of General Harris, of Lieut.-Colonel Montagu's name at the second siege of Seringapatam, where he was universally acknowledged to have greatly distinguished himself. In the attack on Tippoo's position in front of Seringapatam, on the night of the 6th of February, 1792, Major Montagu commanded the artillery with the centre column, under the personal direction of Lord Cornwallis.

When the last war was undertaken against the ruler of Mysore, Lieut.-Colonel Montagu was sent in command of the Bengal Artillery. On the 2nd of May, while watching the effect of the fire in one of the batteries, a round shot shattered his arm near the shoulder, rendering immediate amputation necessary. Notwithstanding the serious nature of the wound, he still caused himself to be carried down to the trenches, and there, by his indomitable spirit and cheerfulness, continued to animate his men. For a few days he seemed in a fair way of recovery; but the chest was contused by the same shot, and mortification set in, which proved fatal on the 8th of May, four days after Seringapatam had fallen. His military talents and experience, as well as his personal qualities, made him respected as well as loved; and few officers have had the good fortune to be as often mentioned with praise as Colonel Montagu, of whom his corps was justly proud.

The dates of his commissions were:—fire-worker, 16th May, 1772; lieutenant, 24th September, 1777; captain-lieutenant, 20th March, 1780; captain, 13th October, 1784; major, 14th September, 1790; lieutenant-colonel, 1st March, 1794.

He married, at Masulipatam, in 1792, a Miss Fleetwood, who with three children survived him.

MAJOR-GENERAL SIR JOHN HORSFORD, K.C.B.

This distinguished officer was born the 2nd of May, 1751. At an early age he was sent to Merchant Taylors' School, and thence to St. John's College, Oxford; but, as it is believed, from a desire to evade entering into the Church, as was intended for him, he enlisted in the East India Company's service in the spring of 1772, and came out to Bengal a private in the artillery, under the name of Rover. The inquiries that were made for him by his family attracted the attention of the commandant, Colonel Pearse, and it was said that, having pointed out an error in a Greek quotation in some papers he was copying out for that officer, the latter was induced to try his identity by suddenly calling out his real name as he was leaving the room. In consequence of his high character as a soldier, and the circumstances under which he had enlisted, it was determined to promote him to a commission, and the necessary sanction having been received, the following letter was addressed to the officer commanding his company:—

"To Captain WATKIN THELWALL,
"Commanding 1st Company Artillery.

"Fort William, 9th March, 1778.

"SIR,

"I am directed by Lieut.-Colonel Pearse to acquaint you that Serjeant Rover, of your company, is in this day's orders appointed a Cadet of Artillery, under the name of John Horsford. He desires that he may proceed to the Presidency immediately, in order to join his corps.

"I am, sir, etc.,
"C. R. DEARE,
"Adjt. Corps of Artillery."

The dates of his commissions were:—lieut.-fireworker, 31st March, 1778; lieutenant, 5th October, 1778; captain, 26th November, 1786; brevet-major, 6th May, 1795, regimental, 1st July, 1801; brevet-lieut.-colonel, 1st January, 1800, regimental, 1st May, 1804; lieut.-colonel commandant, 1st August, 1805; colonel, 25th July, 1810; major-general, 4th June, 1813.

Sir John Horsford served in the second Mysore war, 1791-92, under Lord Cornwallis, in command of the 3rd Company, 2nd Battalion (not now existing), and was present at the siege of Bangalore, battle of Arikera, and operations before Seringapatam. He commanded the artillery during the campaigns under Lord Lake in 1803-4-5, and was honourably mentioned by him in despatches reporting the capture of Aligarh (dated 4th September, 1803), the battle of Delhi (12th September), capture of Agra (18th October), also in G. G. O. of 15th September and 1st October. For the battle and subsequent siege of Deeg, he was mentioned by the Commander-in-Chief both in general orders and in his despatch to the Governor-General, an extract from which will be found in vol. i. chap. viii. p. 263. He also commanded the artillery at the siege of Bhurtpore, which lasted from the 2nd of January till the 11th of April, 1805.

In consequence of the zeal and ability displayed by him in command of the artillery in the field, the Governor-General in Council, in a letter dated 9th October, 1806, recommended him to the Court of Directors for a special allowance while holding that command.

At the siege of Kamonah in August—November, 1807, Lieut.-Colonel Horsford commanded a brigade, but had also the direction of the artillery operations.

On the resignation in May, 1808, of Colonel Nicholas Carnegie, he succeeded to the command of the regiment, which he held till his death. In this position he continued to study, as he had always done, the improvement of the arm of he service he belonged to. At Cawnpore, in October, 1801, he had presented to General Lake, the Commander-in-Chief, a memoir which he had drawn up on the organization of the Bengal Artillery, pointing out its defects and their remedy. He laid before Lord Moira* another memoir on the same subject, in which he showed that the experiences of the French in the late continental wars in Europe had not been unstudied by him. His high character as an officer secured attention, but it was not till after his death that his endeavours bore their fruit. The reorganization of 1817-18 placed the regiment upon a more efficient footing than it had been before.

Major-General Horsford was not sent on service during the Nipál war in 1814-16, but as commandant of artillery he had to carry out, as far as his arm was concerned, the extensive combinations of Lord Moira. And in 1815, when the Order of the Bath was opened for the admission of officers of

* Letter dated Dumdum, 7th June, 1816.

the Indian service, his Majesty was pleased to confer upon him the second class decoration of a Knight Commander, Date of nomination, 7th April, 1815.

He was, by general orders of the 28th of June, 1816, placed upon the staff of the army as an extra major-general; and a few days afterwards, Colonel T. Hardwicke succeeded him as acting-commandant of the Bengal Artillery. But the tenure of Horsford's regimental command only ceased with his life.

In the siege of Háthras, which was taken early in March, 1817, Sir John Horsford was for the last time employed against an enemy. He commanded his own arm, which was prepared upon a larger scale than had before been organized for any military operations in India. The expectations of the Marquis of Hastings were realized by the result. Before leaving the army, Sir John Horsford issued a farewell order to the artillery belonging to it, which is given in its place in this work; and there is something touching in his allusion, in its concluding sentence, to the anticipated close of his military career, so soon to be verified.

"It is a source of great pleasure for the Major-General to reflect that the last period of his service with a corps in which he has long served should be distinguished by events which call forth the admiration of all who witnessed them, and by services which conspicuously increase the credit and the established high character of the regiment of Bengal Artillery.

At Cawnpore, ten days after his return from the army, on the 20th April, 1817, he ended a service in India of forty-five years, "spent in constant and unwearied devotion to his duty, never having, even in sickness, enjoyed the indulgence of one day's furlough or leave of absence from his professional labours." * A sound constitution and strict temperance enabled him to endure what our present nervous temperaments would shrink from. Intellectually, in a scientific knowledge of his profession, in his habits of order and system, and in spotless integrity, he was confessedly unexcelled; regular in the discharge of all his duties, military and domestic, till within

* East India Military Calendar, ii. 310.

a few hours of his death. This was caused by ossification of the heart.

The Marquis of Hastings, in filling up the vacancy, bore the following testimony to his character as a soldier:—

"The Governor-General cannot direct the succession in the regiment of artillery without expressing his deep concern at the loss the Honourable Company has suffered by the death of Major-General Sir John Horsford, K.C.B. The ardent spirit, the science, and the generous zeal of that admirable officer were in no less degree an advantage to the public interest, than an honour to himself. It is consolatory to think that, when sinking under the malady which so early deprived his country of an energy incessantly devoted to her glory, he had the consciousness of having just displayed, with signal triumph, the skill and superiority of the corps which he had so materially contributed to fashion and perfect."

MAJOR-GENERAL THOMAS HARDWICKE

Was appointed a cadet in India in 1778. He served in the 5th Company, 2nd Battalion (now the 4th Battery, 23rd Brigade R.A.), from about 1779, when it was first raised, for the greater part of the time during which he was a subaltern. This company, under the command of Captain C. R. Deare, was sent, in 1781, to serve on the Madras coast during the first Mysore war. Lieutenant Hardwicke returned to Bengal in November, 1783, about four months after the cessation of the siege of Cuddalore, and seven months before the company itself came back. When the second Mysore war broke out, Lieutenant Hardwicke went again with the same company, then commanded by Captain G. F. J. Sampson, to Madras, in April, 1790. He was present at the capture of Errod, in August, 1790; at the battle of Sátiya Mangalam, where Captain Sampson was badly wounded, in September; sieges of Bangalore in March, 1791; of Sávandrug and Utradrug in December; and of Seringapatam in March, 1792.

In October, 1794, being then a captain, he commanded the

1st Company, 3rd Battalion (the present B Battery, 16th Brigade), at the battle of Bitaurah against the Rohillas.

He was on furlough from the 8th of March, 1803, until the 20th of November, 1806. When Major-General Sir John Horsford, in the end of June, 1816, was placed on the staff of the army, Colonel Hardwicke was appointed to act as commandant of the Bengal Artillery; and on the death of that officer the following year, he was confirmed in the post. He went home for two years on sick leave, in February, 1818, during which time Colonel J. D. Sherwood officiated for him; and he finally resigned the command, and went home, 22nd December, 1823. He died at the Lodge, South Lambeth, on the 3rd of March, 1835, aged 79 years.

Major-General Hardwicke was a good average artillery officer, but not in advance of his time, like his predecessor in the command of the Bengal Artillery. His scientific acquirements were extensive, especially in natural history. He had been elected a member of the Asiatic Society of Bengal, on the 29th of September, 1796, and was a vice-president from 1820 to 1822. Only one paper from his pen appears in the Asiatic Researches.* He left large collections of natural history, which were partially worked out by Dr. J. E. Gray; and the "Illustrations of Indian Zoology," a folio volume, with large coloured engravings, was the joint production of Major-General Hardwicke and Dr. Gray.

The dates of his commissions were:—fireworker, 3rd November, 1778; lieutenant, 16th February, 1784; captain, 20th August, 1794; brevet-major, 1st January, 1798; regimental, 26th July, 1804; lieut.-colonel, 21st September, 1804; brevet-colonel, 4th June, 1813; regimental, 21st September, 1817; major-general, 12th August, 1819.

* "Narrative of a Journey to Sirinagar," vol. vi., for 1799.

MAJOR-GENERAL HENRY GRACE

Was a cadet of the year 1778. In March, 1786, when he had been a little more than seven years in the service, he was selected to succeed Captain Charles R. Deare, as adjutant to the Regiment (then called the Brigade) of Artillery; but this office was in April, 1787, given to Lieutenant P. Cullen, of the native infantry, as brigade-major. This anomaly, however, was put an end to the following year, and Lieutenant Grace became brigade-major, which appointment he retained till April, 1806, when, having been promoted to lieut.-colonel, he vacated it. While in the office of brigade major, he compiled the first "Code of Regulations for the Bengal Army," which was published. It contained standing orders from 1786 to the end of 1790. A larger work, comprising a digest of orders going back as far as 1748, had been begun from records collected and indexed under the orders of Colonel Pearse, but was not printed. A second volume of his code appeared in 1799.

On the 3rd of November, 1810, he was sent to Cawnpore, to command the artillery "in the field," and in this position applied himself to improve the condition and working of magazines, which, for want of officers, were left too much in charge of sergeants and warrant officers of the Ordnance Department. He also assisted Government in drawing up a plan for a regular Commissariat Department, to supersede the old method of supplying the wants of the army by contracts with private firms.

Colonel Grace was a scientific officer, but had little suavity of manner, and was unpopular in his relations with others. Lieutenant C. P. Kennedy, who was his staff officer in 1816-17, nevertheless, speaks highly of him, and says that he made all hands under him do their work. He prepared for Lord Hastings the mountain guns and field ordnance required for the Nipál campaign, and subsequently the train for the siege of Háthras, in the early part of 1817. Fever was very prevalent at Cawnpore at the time, and after the labour of getting the siege material ready, both Colonel Grace and Lieutenant Kennedy were compelled to take sick leave; the former to the

Cape of Good Hope, the latter, home. After his return from the Cape, Major-General Grace died at Calcutta, on the 3rd of May, 1820, at the age of 62 years.

The dates of his commissions were :—fireworker, 26th December, 1778; lieutenant, 8th July, 1784; captain, 11th February, 1795; brevet-major, 1st January, 1798; regimental-major, 21st September, 1804; lieut.-colonel, 28th February, 1806; brevet-colonel, 4th June, 1813; regimental-colonel, 1st September, 1818; major-general, 12th August, 1819.

LIEUT.-COLONEL GEORGE CONSTABLE

Joined the artillery as a cadet 5th November, 1781.

His military services were—

Served in the first Mysore war, in 1783, with the army under Sir Eyre Coote.

Proceeded with the force detached to Madras in 1794 for foreign service, but which was not employed.

With the force under Sir James Craig at Anupshahr in 1798. Commanded the artillery sent with General R. Stuart in pursuit of the rebel Nawáb, Vazir Ali, who was followed up to the foot of the Nipál hills, at Tulsipur.

Detached with 7th Regiment, N.I., under Colonel James Morris, which reduced after a stout defence the fort of Dastampúr in Oudh,* May, 1800.

* This place is, as I had conjectured, not in the province of Oudh itself, but within the territory then belonging to Oudh, south of the Ganges. Mr. Irvine, C.S., officiating Collector of Farukhábád, with some trouble, kindly traced out its position for me. It is about 13 miles west of the town of Etah. A local history relates that the Náib of the Talukdár of Farukhábád came to Dastampúr to collect revenue, and that the zamindár of that place opened fire upon the Náib's people, who for several days unsuccessfully tried to reduce the fort. The account goes on to say :—

"At last the Nawáb Sáhib, being at his wits' end (*láchár*), sent for Colonel Máras [Morris] Sáhib with his regiment from the camp [*i.e.* the cantonment of Fatehgahr]. When he reached the place, and had an

Capture of the forts of Sasni and Bijigarh in 1803. At the latter place he was wounded.

In August of the same year, having sufficiently recovered, he left Fatehgarh and joined Lord Lake's army, with which he was present at the capture of Aligarh, battle of Delhi, siege of Agra, and battle of Láswári. He was subsequently with Colonel Monson's force;* but his health failing, he left for Calcutta in May, 1804, whence he proceeded home.

In 1806 he was engaged under the control of the Board of Ordnance, of which the Earl of Moira was then Master-General, in casting composite guns, having a bronze coating over an iron cylinder. The idea was taken from the guns which fell into our hands in Lord Lake's campaign of 1803, and it may perhaps be wondered at, that England should think of borrowing from India an improvement in ordnance. The difficulties, however, of producing a satisfactory kind of weapon built up of different kinds of metal will be better understood at the present day than they were then. The guns were cast in London with the assistance of General Sir Thomas Bloomfield, who furnished materials from Woolwich, and were afterwards surveyed by a committee of field officers of artillery. The thanks of the Board of Ordnance were conveyed to Major Constable. One of the guns thus made was a 3-pounder; its weight was 2 cwt. 3 qrs. 1 lb.—27 lbs. lighter than the 3-pounders then in use, though constructed after the same dimensions, or nearly so. The bore was of iron, which certainly was no improvement. The object of the experiment was to manufacture a gun which should be as light, and less fusible than the bronze ordnance; but it does not appear that any more were constructed upon the same principles. Major Constable also studied under Colonel H. Shrapnel, of the Royal Artillery, the use of the projectile which bears his name, in which, on his return to India, he instructed his own corps. A quantity of shrapnel shells had

interview with the Nawáb, he said, 'Nawáb Sáhib, I will clear out this fort for you in three hours.' In short, the colonel next day caused the fort to be deserted within three hours' time. Many men were lost on both sides. Then the colonel took his leave, and went back to cantonments."—From the "Loh-i-Tárikh," by Mir Bahádur Ali of Chibramau.

been sent out to Bengal by the Court of Directors in 1806.

Lieut.-Colonel Constable retired from the service on the pay of his rank, the 17th of January, 1816.

The dates of his commissions were:—fire-worker, 26th July, 1782; lieutenant, 26th June, 1788; captain-lieutenant, 8th January, 1796; captain, 18th February, 1802; major, 28th February, 1806; lieutenant-colonel, 5th December, 1809.

COLONEL SIR ALEXANDER MACLEOD, Kt., C.B.,

Was a cadet of the year 1783, arrived in Bengal in 1784, and was commissioned as fireworker in the artillery early in 1785. In November, 1790, he was sent with a draft of officers and men, under command of Captain G. Howell, to the army on service in the Madras presidency, and on arrival there was posted, in G. O. of the 3rd of February, 1791, to the 5th Company 2nd Battalion (4-23 R.A.). He served throughout the remainder of the second Mysore war, and after his return to Bengal in 1792 was transferred, on the 30th of August, to the 1st Company 1st Battalion (1-23 R.A.), then at Fort William. He was subsequently transferred to the 3rd Battalion, of which he was appointed the adjutant. After the death, in September, 1805, of Captain C. Hutchinson, Captain Macleod was appointed by Lord Lake to command the post of Tonk Rámpura, which had been held so gallantly by the former during Holkar's irruption into Hindustan. The 2nd Company 2nd Battalion (not now existing) was then at Tonk. He was afterwards transferred to the 3rd Company 1st Battalion (1-22 R.A.), but again held command of the 2nd Company 2nd Battalion, though not posted to it, at the sieges of Kamonah and Ganáori, October—December, 1807. From thence he was moved to the 3rd Company 3rd Battalion (7-23 R.A.); was present with it at the capture, in August, 1809, of the fort of Bhawáni, and retained the command of it till his promotion to major in 1812.

He was then posted to the 3rd Battalion, and in 1814-15

commanded the artillery with Major-General Ochterlony's division in the operations against Amar Singh Thápá, terminating in the capture of the Maláon heights. For these various services, in which his name had been frequently brought forward for favourable notice, he was honoured with the Companionship of the Bath, 3rd of February, 1817. He served at the siege of Háthras in 1817, and at the end of the same year commanded the artillery with the Right Division, under Major-General Sir R. S. Donkin, in the Pindári and Máhrátá war.

After Major-General Grace had left India on account of his health, Lieut.-Colonel Macleod succeeded him in the command of the artillery "in the field," or above Allahabad, which he held till that command, as a distinctive one, was abolished. By the same order (G. G. O. 27th November, 1823) he was appointed brigadier-commandant of the Regiment of Artillery, in the room of Major-General Hardwicke, gone home. When the army was assembled for service against Bhurtpore, he was directed to assume charge of the whole of the artillery employed there, with the rank of a brigadier of the first class (G. O. C. C. 5th December, 1825). As he was not a major-general, he did not obtain the advanced grade in the order of the Bath, but the *London Gazette* published the announcement, dated the 29th of August, 1827, that his Majesty had conferred on him the honour of knighthood.

This was the last of a long series of distinguished services. Sir Alexander Macleod died at Dumdum, on the morning of the 20th of August, 1831, after a brief illness, at the age of 64. He had completed upwards of forty-seven years of uninterrupted service in India. His character as an officer, and his kindness of heart, caused him to be universally respected and liked, not only in the regiment, but by the army at large.

The dates of his commissions were :—fireworker, 18th April, 1785; lieutenant, 7th April, 1793; brevet-captain, 8th January, 1798; captain-lieutenant, 21st February, 1802; captain, 21st September, 1804; brevet-major, 25th July, 1810; regimental, 1st March, 1812; lieut.-colonel, 15th February, 1818; lieut.-colonel-commandant, 1st May, 1824; colonel, 5th June, 1829.

COLONEL GERVAISE PENNINGTON, C.B.,

Came out to India in the year 1783 as an infantry cadet, but was commissioned a fireworker in the artillery the same year.

He served at the taking of Seringapatam in 1799, and after his return to Bengal was appointed aide-de-camp to Major-General George Deare, Bengal Artillery, who was commanding a division on the staff of the army. From this, in 1803, he was appointed officiating commissary of ordnance at Chunár, and was sent up to Bhurtpore, at the first siege of that place, on account of the paucity of officers there. He arrived in camp on the 30th of January, 1835, and was present at the last two assaults, distinguishing himself in the trenches.

Next year, Captain Clement Brown having been appointed commissary of ordnance in Fort William, the command of the "experimental brigade," as the troop of horse artillery was then termed, was given to Captain Pennington, and he remained in this branch as long afterwards as he was in the service. To him, in its earlier years, and to the example he set those who, as the Bengal Horse Artillery was augmented, were successively appointed to the command of batteries, may be in a great measure attributed their efficiency. The system of drill and field manœuvres which he introduced was the groundwork of that which long continued in use in the regiment.

In October, 1814, Captain Pennington proceeded with his own, the 1st Troop, and the 3rd Troop of Horse Artillery, into the Dera Dun, with Major-General Sir R. Gillespie's Division. The fort of Kalanga was attacked, and it was at one of the guns of the 1st Troop, under Lieutenant C. P. Kennedy, within a few feet of the entrance, that the general was killed. Captain Pennington and his young subaltern were several times honourably mentioned in despatches. After the evacuation of Kalanga, the horse artillery returned to Meerut, not being adapted for mountain warfare.

On promotion to the rank of major, he was retained in command of the mounted branch, though as yet horse artillery officers were still borne on the strength of their foot artillery companies and battalions. He was present at the siege of

Háthras, and commanded the Horse Artillery with the Centre Division, in the Pindári war of 1817-18.

In January, 1825, he went home on furlough, and in the end of the same year married Jane, second daughter of J. P. Grant, Esq., of Rothiemurchus. He was subsequently, for his services, nominated a Companion of the Bath. Date of nomination, 26th of September, 1831.

He went home again on account of his health, and died at Basingstoke, on the 2nd of July, 1835. Captain Gervaise Pennington, who came out in the year 1816, and died at Meerut, 13th October, 1835, when in command of the 3rd Troop, 1st Brigade H.A., was a nephew of his.

The dates of Colonel Pennington's commissions were as follows:—fireworker, 12th May, 1785; lieutenant, 12th June, 1793; brevet-captain, 8th January, 1798; captain-lieutenant, 23rd February, 1802; captain, 21st September, 1804; brevet-major, 25th July, 1810; regimental-major, 17th February, 1815; lieutenant-colonel, 1st September, 1818; lieutenant-colonel commandant,* 1st May, 1824; colonel in the army, 5th June, 1829.

MAJOR-GENERAL CLEMENT † BROWN, C.B.

This officer was descended from an Irish family. He was appointed to a cadetship in 1783, arrived in Bengal in July, 1784, and was gazetted an ensign of infantry on that establishment on the 3rd of February, 1785. Owing to the great deficiency of officers in the Bengal Artillery about this time,

* In other words, colonel regimentally; but the rank, according to the system which then (and for a long time afterwards) was in force, was not given till the officers of H.M.'s army of similar standing had been promoted. Lieutenant-colonels commandant received therefore the emoluments, without the rank, of a regimental full colonel. The original order on this head (somewhat modified afterwards) was contained in letter from the Court of Directors, dated 23rd December, 1806, published in G. G. O. 28th July, 1807.

† This name is often spelt with a final *s*, but it is believed that the above is the correct way.

a number were transferred from the infantry in 1788 and the two following years. Ensign Brown, with three others,* was commissioned in the artillery in the early part of 1789, and posted to the 2nd Battalion, he falling to the 1st Company (3-23 R.A.) This battalion was sent to the Madras coast on service in February, 1790, and Lieutenant Brown was employed throughout the whole of the war with Tippoo Sultán, terminating in his submission, in February, 1792, to Lord Cornwallis, beneath the walls of Seringapatam.

After his return to Bengal, Lieutenant Brown was transferred to the 1st Company 3rd Battalion (B-16 R.A.), and was probably appointed adjutant to the battalion, for we find him acting in that capacity with the artillery employed against the Rohillas in October, 1794. In the battle of Bitaurah, when the misconduct of the cavalry caused so much loss to the right of the line, Lieutenant Brown was stationed with the artillery in the centre, and his exertions in bringing his guns to bear with grape upon the masses of the enemy were most conspicuous. For about fifteen minutes a bloody hand-to-hand contest was maintained upon the right, which, though thrown into confusion, was not driven back. The conduct of the artillery and infantry, both native and European, was excellent.

In January, 1800, Lord Wellesley being desirous of raising, experimentally, a troop of horse artillery, Captain Brown was selected by him to superintend its formation. A small portion of it, two guns and nine horses, was despatched to Egypt in 1801, and joined General Baird at Kosseir about the beginning of July. In the march through the desert to Ghenneh, however, he lost his horses from the want of water, and the guns were taken on by camels. The force saw no fighting, and on the 4th of August, 1802, Captain Brown and his men were back again in Bengal. He was then appointed adjutant of the 1st Battalion, but reverted to the horse artillery in 1804.

The "experimental brigade" was not employed in the

* Their names were George Jones, William Winbolt, and William Horatio Green. They all served in the Mysore war of 1790-92. See vol. i. pp. 106, 107.

first campaign against the Máhrátás in 1803, in time to share in any of the actions; it joined General Lake in December, while head-quarters were halted at Biána, near Agra. Thence it accompanied the army in its movements * until it returned to Cawnpore in the following June. In September Captain Brown marched up to Agra, where the army was being again assembled for service. It was at the village of Áring (October 7th, 1804) that the horse artillery guns were first unlimbered for action in front of an enemy. Holkar's horse were in line with their right on the village, and their left extending northwards, and Lake had moved out to surprise them; but they were prepared. The despatch says:—

"On perceiving this, I ordered the horse artillery under Captain Brown to advance, and on their commencing a brisk cannonade on the enemy's right, these were immediately thrown into confusion, and commenced a precipitate flight. The cavalry moved forward against the left of their position, and did some execution with their gallopers, but the rapidity of the enemy's flight rendered it impossible to effect a charge."

The subsequent events of this campaign have already been detailed. Horse artillery was then but a new arm, virtually, in warfare, but the services of this battery in General Lake's march after Holkar, and in the pursuit of Amir Khán through Rohilkhand, show the care that Captain Brown must have bestowed upon his horses, men, and *matériel*, and are a sufficient testimony to his excellence as a horse artillery officer. At the battle of Afzalgarh, he was destined again to have his arm compromised by the injudicious handling of the cavalry, which were driven back through his guns; and the gunners, unable to fire into the mixed body of friends and foes, were loading their pistols and preparing for self-defence the best way they could, when Captain Deare's gallant charge relieved them from their critical position (vol. i. p. 279). It should be mentioned that the depth of the ground in front of the line was said to have been limited for a cavalry charge by the steep banks of a small stream, concealed by long grass, which did not hinder an effective charge from the flank along the front.

* See route in note D in the appendix to chap. vii. vol. i.

On the 23rd of January, 1806, Captain Brown was appointed commissary of ordnance, and made over the command of his troop to Captain G. Pennington. He continued in that department, which was one of considerable emolument in those days, until, in 1810, he went home on furlough. He returned in January, 1812, with Sir George Nugent, who was coming out as commander-in-chief, and who appointed him one of his aides-de-camp. When the gun-carriage agency at Allahabad was established, Major Brown was appointed (G. G. O. 23rd June, 1814) to the charge of it.* The agency was shortly afterwards transferred to Fatehgarh, where it now is, and Major Brown retained the appointment until, in 1824, he was promoted to the rank of lieut.-colonel-commandant.

In the reorganization of the regiment of this year, Colonel Brown was appointed to the command of the 3rd Brigade of Horse Artillery, and the exertions he made to increase the efficiency of this branch showed that the long time which had elapsed since he belonged to it had not unfitted him for the command, or lessened his love for it. Much difference of opinion existed at this time as to the proper calibre of ordnance for horse artillery batteries. The story will be told in its proper place. In the discussion of nine *versus* six-pounders, though the nature of the service in which in Lord Lake's time he had been employed might have induced a different conclusion, his desire to render horse artillery batteries something more than mere "brigades of gallopers" led to his advocating the use of the heavier pieces. To his exertions it was due that for a few years they were ordered to be armed with-nine pounder guns and the heavier calibred field howitzers.

In the early part of the Burmese war, the commander-in-chief, Sir Edward Paget, who had some thoughts of taking the command of the army before Rangoon, selected Colonel Brown to go with him as senior artillery officer, but the arrangement was never carried out. In January, 1825, Colonel Pennington having gone home on furlough, the command of the horse artillery devolved upon him as next senior officer,

* The salary allotted for this appointment was Sonat rupees 1254-1-0, in addition to regimental pay.

and he was afterwards (G. G. O. 3rd October, 1828) nominated a brigadier for the purpose of inspecting the different brigades and detached troops. He had commanded this arm at the second siege of Bhurtpore, which he had also seen attacked in 1805, and for this and his other services he was (26th December, 1826) nominated a Companion of the Bath.

By G. G. O. 23rd of September, 1831, he was appointed commandant of the regiment, on the death of Sir Alexander Macleod, and held that post till, in August, 1836, he was placed upon the staff of the army as a brigadier-general in command of the Benares division. His promotion to the rank of major-general soon followed, but the period of his services was drawing to a close. He died at Benares, on the 24th of April, 1838, in the seventy-second year of his age. He was warm-hearted and impulsive in temperament, like most of his countrymen; and, though somewhat eccentric in his bounty in later years, ever liberal with his purse, whether it were to subscribe fifteen hundred rupees for building a racquet court,* or to back a gunner against a dragoon to ride a horse between the stables of their respective lines.

The house he lived in at Meerut, the same one occupied at the time of the Mutiny by the assistant adjutant-general of artillery, bordered upon the road which Major-General Sir F. Wheler, commanding at that station a few years ago, named after him. There are not many in the service now to whom the designation would recall the gallant old soldier who in his time had done not a little to enhance the reputation of the corps he belonged to.

The dates of his commissions in the artillery were:—fire-worker, 1st February, 1789; lieutenant, 1st December, 1794;

* The one in the artillery lines at Meerut. But the quota given by a second lieutenant (G. Campbell) was two hundred and fifty, which was not out of proportion with that of the brigadier.

The wager referred to was, I believe, that a man from the horse artillery was to be provided with a dragoon horse, and the dragoon with an artillery horse; each man was to ride *away from* his own stables to those of the other corps, and, if he could get there, to ride back again. Troop horses were, many of them, notably vicious, especially in the gun teams, and of course the quietest was not selected. I do not know the result of the wager, or if it ever came off. The stake proposed was, I think, fifty gold mohurs.

brevet-captain, 8th January, 1798; captain-lieutenant, 25th December, 1802; captain, 28th February, 1806; brevet-major, 25th July, 1810; regimental major, 21st April, 1817; lieutenant-colonel, 2nd August, 1819; lieutenant-colonel-commandant, 1st May, 1824; colonel, 5th June, 1829; major-general, 10th January, 1837.

LIEUT.-COLONEL SIR GEORGE EVEREST, K.C.B.

Born at Gwern Vale, Brecon, on the 4th of July, 1790. Although not yet of the full age when he passed his final examination from Woolwich, it was with such success that he was pronounced fit for a commission, and was gazetted accordingly to the Bengal Artillery as a lieutenant * on his arrival in India. Date of rank, 4th April, 1806.

After about a year's duty he was sent to the island of Java, and while there was employed by Sir Stamford Raffles to execute a reconnoitring survey of the country. After his return to India he was appointed, in conjunction with Lieutenant R. B. Fergusson, of the Rámgarh battalion (G. G. O. 21st October, 1817), to survey the proposed line for telegraphic communication between the presidency and Chunár; and two years after was appointed the chief assistant to Colonel Lambton, the celebrated founder of the great trigonometrical survey of India, whom he succeeded in 1823, as superintendent of that survey. Captain Everest now proceeded to concentrate the resources at his command for the extension of the great arc series, and succeeded, in spite of formidable difficulties, in carrying the measurement at length to the latitude of 24°.

In the year 1826, his health, which had been seriously affected by the laborious duties of his office, compelled him to go home, where he made himself acquainted with the English ordnance survey system, and with all the modern improve-

* All the cadets of this season were promoted at once to the rank of lieutenant, to fill existing vacancies.

ments in geodesical matters. While still in England, in 1829, he was appointed Surveyor-General of India, in addition to the post of Superintendent of the Great Trigonometrical Survey. He left England in June, 1830, liberally provided, by the munificence of the Honourable Court of Directors, with instruments and apparatus superior to any then in the world, and arrived in India in October. The following account of the measurement of a verification base, near Calcutta, on the Barrackpore road, shows how carefully all his arrangements were made, and how efficient a staff of assistants he had under him:—

"The measurement of this base commenced on the 23rd of November, 1831, and ended on the 31st of January, 1832, an interval of fifty-eight days, of which thirteen may be set down as holidays, so that the actual time employed was about forty-five days. The length of the ground measured upon an average was 750 feet, or twelve sets of bars; but towards the conclusion, so systematic had become the operations, that eighteen, twenty, and once twenty-four sets (that is, 1512 feet) were measured in one day, which is double what was effected on the Irish survey. This was chiefly attributable to the number and experience of the officers employed."

The length of this base was nearly 34,000 feet, and the difference, on remeasurement, was so small, that it would only have amounted to 125 feet in the diameter of the globe.

In 1832, after an interval of seven years, the great arc was recommenced, and the work was terminated in 1841 with the measurement of the Beder base line, executed by his chief assistant and astronomer, Captain (now Major-General Sir) Andrew Scott Waugh. This arc extended from the 30th to the 8th degree of latitude, from the Himaláyas to Cape Comorin, down the centre of the Indian peninsula. The Asiatic Society of Bengal, in nominating him one of their honorary members, thus notice this splendid achievement:—

"By the light it throws on researches into the figure and dimensions of the earth, it forms one of the most valuable contributions to that branch of science which we possess, whilst at the same time it constitutes a foundation for the geography of Northern India, the integrity of which must for ever stand unquestioned."

In 1837 Lieut.-Colonel Everest had again suffered so much in health, that it was thought that he would not be able to retain his post, and Major Jervis, of the Bombay Engineers, was appointed provisionally to succeed him. This officer submitted to the Royal Society certain proposals for the extension of science, and the improvement of the geography of India; and that body (of which Lieut.-Colonel Everest was a member) brought the subject to the notice of the Court of Directors. Lieut.-Colonel Everest took exception to this, which he considered tantamount to a want of confidence in himself, as he said that the address, signed by the President and thirty-eight members of the Royal Society, to the Directors, called upon them "in language little short of peremptory, to repose confidence in, and delegate power to, Major Jervis." The correspondence on this subject, embodied in a series of letters addressed to H.R.H. the Duke of Sussex, as President of the Royal Society, was published by Lieut.-Colonel Everest, in London, in 1839.

It was not, however, until the 16th of December, 1843, that he retired from the service. The retirement or death of the more distinguished servants of the Honourable East India Company was generally announced in official despatches and orders, and the following was published upon this occasion:—

"GENERAL ORDERS BY THE PRESIDENT IN COUNCIL.

"Fort William, 22nd December, 1843.

"No. 284. The Honourable the President in Council has much pleasure in publishing the following paragraphs of a letter from the Honourable the Court of Directors to the Government of Bengal, No. 19, dated the 3rd of May, 1843:—

"Para. 1. The announcement of the intended return to Europe of Lieut.-Colonel Everest, and his consequent vacation of the office of Surveyor-General and Superintendent of the Great Trigonometrical Survey of India, affords us an opportunity of which we readily avail ourselves, of expressing the high sense we entertain of the scientific acquirements of that officer, and of the ability and zeal which he has displayed in the discharge of the arduous duties entrusted to him.

"2. With the measurement of an arc of the meridian of unprecedented magnitude, now finally completed, and with the Great

Trigonometrical Survey, the most important portion of which is now in rapid progress towards completion, Lieut.-Colonel Everest has been prominently connected almost from their commencement, and for the last twenty years they have been under his exclusive management and superintendence.

"3. The able manner in which he has conducted these important and scientific works has frequently elicited our approbation, and cannot fail to cause his name to be conspicuously and intimately associated with the progress of scientific inquiry."

For these services Colonel Everest was, some years after his return home, made a Companion of the Civil Division of the Order of the Bath, and later, a Knight Commander. He died at No. 10, Westbourne Street, London, on the 1st of December, 1866.

The dates of his commissions were :—lieutenant, 4th April, 1806; captain-lieutenant, 25th September, 1817; captain, 1st September, 1818; major, 25th July, 1832; lieut.-colonel, 7th March, 1838.

A part of this sketch has been taken from the notice of Colonel Everest in the Annual Report of the Royal Asiatic Society, vol. iii. New Series, for 1867, p. xvi.

INDEX TO
NAMES OF BENGAL ARTILLERY OFFICERS.

Abbott, A., second siege of Bhurtpore, xiv.
Abbott, J., second siege of Bhurtpore, xiv.
Adamson, B., killed at or near Patna, i.
Addison, E., first Mysore war, iii.
Alexander, J., second siege of Bhurtpore, xiv.
Allen, Sub-Lieutenant, first Burmese war, xiii.
Anderson, W., second siege of Bhurtpore, xiv.
Archer, C., dies on passage to Java, ix.

Backhouse, J. B., second siege of Bhurtpore, xiv.
Baillie, A. W., first Rohilla war; Bombay detachment; commands the force, ii.
Baillie, E. J., twice honourably mentioned for service near Benares, ii.
Baker, E., second Rohilla war; killed, v.
Baker, H. C., Mauritius, ix. Pindári war, xii.
Baker, O., Nipál war, second campaign, x.
Balfour, H., Bijigarh, ii. Second Mysore war, iv. Third Mysore war, vi.
Barton, J., first Mysore war, iii. Second Mysore war; siege of Bangalore, iv. Ceylon, v.
Battine, W., Nipál war, 2nd Division, x. Háthras, xi. Pindári, xii.
Bazely, F. R., second siege of Bhurtpore, xiv.
Beaghan, F., Máhrátá war; Agra, dies of his wound, viii.
Bedingfield, R. G., first Burmese war; Assam, xiii. Note regarding, xiii. App. A.
Begbie, A. P., first Burmese war, xiii.
Bell, C. H., Pindári war, xii.
Bell, W., Java, ix. Nipál war; Kamáon, x. Pindári war; Dhamoni, xii. Second siege of Bhurtpore, xiv.

Best, R., Sásni; Bijigarh; Kachaura; Máhrátá war, Aligarh, vii.
Biddulph, E., first Burmese war, xiii.
Biggs, J. A., Nipál war, second campaign, x. Pindári war; Jáwad, xii. Second siege of Bhurtpore, xiv.
Bingley, T. W., second siege of Bhurtpore, xiv.
Blake, Errol, first Burmese war, xiii.
Blake, G., Pindári war, Nágpur, xii. Second siege of Bhurtpore, xiv.
Blundell, W., first Mysore war, iii.
Boileau, F. B., second siege of Bhurtpore, xiv.
Boileau, J. P., Java H.A., ix. Háthras, xi. Pindári war, xii.
Boyce, B., Bombay detachment, ii.
Boyle, H. D., third Mysore war, vi. Sásni; killed, vii.
Brady, J., first Burmese war, xiii.
Brind, F., first Burmese war; Assam, xiii.
Briscoe, J. J., second Mysore war; siege of Bangalore, iv.
Broadbridge, J., Condore; Masulipatam, siege of, i.
Brooke, G., Bandelkhand, ix. Nipál war, 3rd Division, x. Háthras, xi. Pindári war, xii. Second siege of Bhurtpore, xiv.
Brooke, J. H., third Mysore war, vi. Hirápur; Rajáoli; Ajigarh, ix. Nipál war, 2nd Division, x. Háthras, xi Pindári war, xii.
Brown, Birnie, first Burmese war, xiii.
Brown, Clement, second Mysore war; Sátiya Mangalam; siege of Bangalore; Utradrug, iv. Second Rohilla war, v. Egypt with experimental H.A., vi. Máhrátá war, first campaign, vii. Second campaign; pursuit of Holkar and battle of Fatehgarh; sieges of Deeg and Bhurtpore, viii. Second siege of Bhurtpore, xiv
Brown, E., first Mysore war, iii.
Brown, J., killed at or near Patna, i.
Brown, R., third Mysore war, vi.
Browne, M. W., Máhrátá war; Aligarh, Delhi, Agra, vii. Battle of Deeg, sieges of Deeg and Bhurtpore, viii.
Bruce, R., Dhálingkot; Bombay detachment; Bijigarh, ii. Second Rohilla war, v.
Bruce, W., first Mysore war, iii.
Buchan, A., second Mysore war; siege of Bangalore; killed at Seringapatam, iv.

Buck, N., killed at Masimpur, i.
Burlton, P. B., first Burmese war; Assam, xiii. Note relative to, xiii. App. A.
Butler, D., first Rohilla war, ii.
Butler, E. W., second Mysore war; Seringapatam, iv. Máhrátá War; Aligarh, Delhi, Agra, vii. Battle and siege of Deeg, viii. Java; Jojokarta, ix. Háthras, xi. Pindári war, xii.

Caldwell (G.C.B.), A., third Mysore war; Málávelli, Seringapatam, vi. Java; Weltervreeden, Cornelis, ix.
Cameron, A., Java; Jojokarta, ix.
Campbell, C. Hay, Hirapur, Rajáoli, Ajigarh, ix. Háthras, xi.
Campbell, G., first Burmese war, xiii.
Campbell, G. N. O., Nipál war, 2nd Division, x. Pindári war, xii. Second siege of Bhurtpore, xiv.
Campbell, Gunner, gallant conduct at Dhálra, vii.
Campbell, Surgeon C., Pindári war, Centre Division, xii.
Cardew, Ambr., first Burmese war; Árákán, xiii.
Carne, J. C., Háthras, xi. Pindári war; Dhámoni, Mandalah, Sátanwári, xii.
Carnegie, N., first Mysore war, iii.
Cartwright, J., gallant conduct at Maláon, Nipál war, second campaign, x. Háthras, xi.
Cautley, P. T., second siege of Bhurtpore, xiv.
Chadwick, T., Pindári war, xii.
Chesney, C. C., Nipál war, 2nd Division, x.
Clarke, C., first Rohilla war, ii.
Clarke, E., second Mysore war, iv. Ceylon, v. Seringapatam, vi.
Clerk, H., second siege of Bhurtpore, xiv.
Coates, H., Bombay detachment, ii.
Constable, G., first Mysore war, iii. Sásni; Bijigarh; Kachaura; Máhrátá war; Aligarh, Agra, Delhi, vii.
Cookson, G. J., second siege of Bhurtpore, xiv.
Coulthard, S., Háthras, xi. Pindári war; Dhámoni, Garhákota, xii.
Counsell, W. E. J., Nipál war, second campaign, x. Pindári war, xii. First Burmese war, xiii.
Cranch, P., second Mysore war, iv.
Crawfurd, G. R., Pindári war; Siuni, Chándá, xii.

Crommelin, J. D., Háthras, xi. Pindári war; Dhámoni, Mandalah, Garhákota, xii.
Cross, Corporal J., his gallant conduct at Zamina, Karáwal, and Dhálra, vii.
Croxton, T., Háthras, xi. Pindári war, reserve, xii.
Cruikshank, K., Nipál war, 3rd Division, x.
Cullen, J., second siege of Bhurtpore, xiv.
Cummings, Stuart, first Rohilla war, ii.
Curphey, W., Háthras, xi. Second siege of Bhurtpore, xiv.
Curtis, J., Pindári war, xii.

Daniel, J. H., first Burmese war, xiii.
Dashwood, F., second siege of Bhurtpore, xiv.
Deare, C. R., first Mysore war, iii. Second Mysore war; killed at Sátiya Mangalam, iv.
Deare, G., first Rohilla war, ii. Máhrátá war; commands at Mirzapur, vii.
Debrett, J. E., Pindári war; Asirgárh, xii.
Decker, Ardean, killed at or near Patna, i.
Delafosse, H., Java; expedition to Palimbang; Celebes, ix. Háthras, xi. Pindári war, xii.
Dickson, R. C., Nipál war, second campaign, x. Háthras, xi. Pindári war, xii. Second siege of Bhurtpore, xiv.
Dixon, C. G., Nipál war, 2nd Division, x. Háthras, xi. Pindári war, xii.
Douglas, H., first Mysore war, iii. Second Mysore war, iv.
Douglas, R., Ceylon, v. Third Mysore war, vi.
Dowell, T., second Mysore war, iv. Bundelkhand, vii.
Doxat, Benj., first Rohilla war, ii.
D'Oyly, T., Pindári war; Mandalah, Asirgárh, xii.
Drummond, J. P., second Mysore war, iv. Third Mysore war; Egypt, vi. Máhrátá war; Guzarát, vii.
Duff, Patrick, Buxár, i. Second Mysore war; sent to command; his improvements in the siege train, iv.
Duncan, F. K., second siege of Bhurtpore, xiv.
Dundas, J. F., Java; Cornelis, ix.
Dunn, A., second Mysore war, iv. Second Rohilla war, v. Third Mysore war, vi. Máhrátá war; Guzarát, vii. Joins at Bhurtpore with Bombay force, viii.
Dunn, W., first Mysore war, iii.
Dyke, G. Hart, first Burmese war; Árákán, xiii.

Edwards, J., second siege of Bhurtpore, xiv.
Elliott, W., first Mysore war, iii.
Ellis, G., second siege of Bhurtpore, xiv.
Elwood, T. M., second Mysore war, iv.
Ewart, J., Mauritius, ix.
Ewart, D., second siege of Bhurtpore, xiv.
Exshaw, J. R., first Mysore war, iii. Second Mysore war, iv.

Faithful, E., Máhrátá war; Bárabatti, vii. Gohad, ix.
Faithful, H., Java; Cornelis, ix.
Farnabie, L. M., Java; killed at Cornelis, ix.
Farrington, J. J., Java; Cornelis, Celebes, ix. Pindári war, xii. Second siege of Bhurtpore, xiv.
Feade, W., second Mysore war, iv. Máhrátá war; killed at Bela, vii.
Fenning, S. W., first Burmese war; Árákán, xiii.
Ferris, J., Gohad; Ajigarh, ix.
Flemyng, W., first Mysore war, iii. Egypt, vi.
Forbes, Alex., Bombay detachment, ii.
Fordyce, J., first Burmese war; Árákán, xiii.
Fordyce, T. D., Mauritius, ix. Háthras, xi.
Fortnam, T., second Mysore war, iv.
Fraser, A., China, vii. Háthras, xi. Pindári war, xii.
Frith, W. H. L., Máhrátá war; first siege of Bhurtpore, viii.
Fuller, G., China, vii. Kálinjar, ix.

Garbett, H., second siege of Bhurtpore, xiv.
Garrett, W. T., second siege of Bhurtpore, xiv.
Geddes, W., Nipál war, second campaign, x. Pindári war, vii.
Gillespie, J., Bijigarh, ii.
Glass, A., second Mysore war, iv. Third Mysore war, vi.
Glass, A. or J., Major Camac's detachment, ii.
Gordon, J., Sásni; killed at Bijigarh, vii.
Gowan, E. P., Kálinjar, ix. Nipál war, 2nd Division, x. Háthras, xi. Pindári war, xii.
Gowan, G. E., Java H.A., ix. Pindári war, xii.
Gowing, Rayner, Máhrátá war; battle and siege of Deeg; killed at the siege of Bhurtpore, viii.
Graham, A., Mauritius, ix.

Graham (C.B.), C., Nipál War, 3rd Division, x. Pindári war, reserve, xii. First Burmese war, xiii.
Graham, E., Ceylon, v. Third Mysore war, vi.
Graham, G. T., first Burmese war, xiii.
Gramshaw, R. M. O., Hirapur; Rajáoli; Ajigarh, ix.
Grand, J. E., operations near Benares, ii.
Grant, C., first Burmese war, xiii.
Green, Christ., first Mysore war, iii. Reprimanded for heading a written remonstrance from certain officers passed over in the roster for service, iv. Second Rohilla war, v.
Greene, J. R., first Burmese war; Árákán; dies of fever, xiii.
Greene, T., first Mysore war, iii. Second Mysore war, iv. Sásni; Bijigarh; Kachaura; Máhrátá war; Aligarh, Delhi, Agra, vii.
Green, W. H., second Mysore war, iv. Third Mysore war, vi. Kálinjar, ix.
Groat, D., first Mysore war, iv.
Grote, F., second siege of Bhurtpore, xiv.
Grove, L. R., killed at the siege of Deeg, viii.

Hall, E., Nipál war, 2nd Division, x. First Burmese war; Árákán, xiii.
Hamilton, R., Bombay detachment, ii.
Hardwicke, T., first Mysore war, iii. Second Mysore war, iv. Second Rohilla war, v.
Harris, C., Kamonah; Ganáori; Bhawáni; Java; Celebes, ix.
Harris, J., died with the Bombay detachment, ii.
Harris, W., first Mysore war, iii.
Hart, R., Major Camac's detachment, ii.
Hastie, Assistant-Surgeon, Pindári war; Nágpur subsidiary force, xii.
Hay, S. S., third Mysore war, vi. Bombay; Máhrátá war; Agra; Laswári; capture of Tonk Rámpura, vii. Battle and siege of Deeg; first siege of Bhurtpore, viii.
Hele, J. S., Java; expedition to Palimbang, ix. Háthras, xi.
Henderson, Surgeon, second Mysore war, iv.
Herbert, P., first Mysore war, iii.
Hetzler (C.B.), R., Ceylon, v. Third Mysore war, vi. Máhrátá war; Bárabatti, vii. Kamonah; Ganáori, ix. Pindári war; Dhámoni, Mandalah, Garhákota, xii. Second siege of Bhurtpore, xiv.

Hill, Justly, operations near Benares, ii.
Hill, T., second Mysore war, iv.
Hind, A., second Mysore war, iv. China, vii. First siege of Bhurtpore, viii.
Hislop, J., Corporal, his gallant conduct at Zamina, Káráwal, and Dhálra, vii.
Hockler, G. F., killed at or near Patna, i.
Hodgson, W. E. J., second siege of Bhurtpore, xiv.
Holland, T., first Mysore war, iii.
Hollingbury, J., first Mysore war; dies at Madras, iii.
Hopper, W., second Rohilla war, v. Bandelkhand, vii. Gohad; Chumir, ix.
Horsburgh, J., first Mysore war, iii. Second Mysore war; wounded at Sátiya Mangalam; dies at Bangalore, iv.
Horsford (K.C.B.), Sir J., second Mysore war, iv. Máhrátá war; Aligarh, Delhi, Agra, vii. Battle and siege of Deeg; first siege of Bhurtpore, viii. Kamonah; Ganáori, ix. Háthras; his death and character, xi.
Horsford, R., second siege of Bhurtpore, xiv.
Hotham, J., first Burmese war; Árákán, xiii.
Howell, G., second Mysore war, iv.
Hudson, P., Matross, his gallant conduct at Dhálra, vii.
Hughes, H. P., Pindári war, xii. Second siege of Bhurtpore, xiv.
Humphreys, R., sent to Ceylon; taken prisoner and put to death, v.
Hussey, Vere W., first Rohilla war, ii. First Mysore war; badly wounded, iii.
Hutchinson, C., Máhrátá war; Aligarh, Delhi, Agra; detached with Colonel Monson; conduct during the retreat; left at Tonk; captures Khatáoli, Zamina, Báhmangáon, Karáwal, Dhálra; dies, vii. His conduct at Bencoolen, vii. App. E.
Huthwaite, E., Nipál war, first campaign; Colonel Gregory's detachment; second campaign, General Nicol's column, x. Pindári war, xii. First Burmese war; Assam, xiii. Second siege of Bhurtpore, xiv.
Hyde, J. C., Háthras, xi. Pindári war, xii. Second siege of Bhurtpore, xiv.

Jennings, W., comes to Bengal; Plassy; Giria; Udwah Nálá;

Patna; is left in command of the army; his judgment in quelling a mutiny; promoted for good service; death, i.
Johnson, J., Pindári war, xii. Second siege of Bhurtpore, xiv.
Johnston, G., second Mysore war; succeeds Conan as brigade-major, iv.
Johnstone, T., Gunner, gallant conduct of, and killed at Dhálra, vii.
Jones, G., second Mysore war; dies at Nagore (Negapatam), iv.

Kaylor, G. F., killed at the battle of Giria, i.
Kempe, R. R., Nipál war, second campaign, x. Pindári war, xii.
Kemptz, J., Bombay detachment, ii.
Kennedy, C. P., Nipál war, 2nd Division; Kalanga, x.
Kinch, J., killed at or near Patna, i.
Kirby, J. S., Pindári war; Mandalah, xii. First Burmese war; Árákán, xiii.

Lane, J. T., first Burmese war; Assam, xiii.
Lawrence, H. M., first Burmese war; Árákán, xiii.
Lawrence, L., Háthras, xi.
Lawrenson, G. S., first Burmese war, xiii.
Legertwood, A., Major Popham's detachment; Major Camac's detachment, ii.
Levey, Matross W., Nipál war; his gallantry at Parsa, x.
Lewin, W. C. J., first Burmese war; Árákán, xiii.
Lindsay, A., Kamonah; Ganàori; Gohad, ix. Nipál war, second campaign (wounded), x. Háthras, xi. Pindári war; Dhámoni; Maudalah, xii.
Lumsden (C.B.), T., Háthras, xi. Pindári war, centre Division, xii. First Burmese war; wounded at Napádi, xiii.
Luxford, J. B. B., Nipál war; killed at Kalanga, xii.
Lyon, Hugh, first Rohilla war, ii.
Lyons, Theod., Nipál war, 2nd Division, x.

Macalister, D., Háthras, xi. Pindári war, xii.
Macdermot, M., first Mysore war, ii.
Macdonald, A., first Mysore war, ii.
Macdonald, J. H., first Burmese war, xiii.

Macdowell, Surgeon J., Pindári war, xii.
Macgregor, R. G., first Burmese war, xiii. Second siege of Bhurtpore, xiv.
Mackay, D. Æ., second siege of Bhurtpore, xiv.
Macklewaine, E., Major Camac's detachment, ii.
Maclean, G., second siege of Bhurtpore, xiv.
Maclean, J., Bombay detachment, ii.
Macleod (Kt., C.B.), Sir A., second Mysore war, iv. Kamonah; Ganáori; Bhawáni, ix. Nipál war, 3rd Division, x. Háthras, xi. Pindári war, xii.
Macphee, F., first Mysore war; died on service, iii.
Macpherson, D., second Mysore war; dies of wounds received at Arikera, iv.
Maidman, W. R., second siege of Bhurtpore, xiv.
Marshall, T., Aligarh, ix. Nipál war; Colonel Gregory's detachment, x.
Mason (C.B.), G., Bhawáni, ix. Háthras, xi. Pindári war, xii.
Mason, Kender, Nipál war, 3rd Division, x.
Matheson, P. G., Nipál war, 1st Division; his gallant conduct at Parsa, x. Pindári war; blows open the gate at Jáwad, xii.
Matthews, A. N., second Mysore war, iv. Sásni; Bijigarh; Máhrátá war, Aligarh, Delhi (lost a leg); appointed commissary of ordnance, vii.
Mayaffre, J., Major Popham's detachment; rashly attacks Rámnagar, and is killed, ii.
Maynard, F., second Rohilla war, v.
Maud, J. D., first Mysore war; dies on service, iii.
McCulloch, Assistant-Surgeon, second Mysore war, iv.
McDowell, J., Nipál war, 4th Division; wounded at Jitgarh, x. Pindári war; Chándá, xii.
McIntyre, J., second Rohilla war, v.
McLeod, D., Ajigarh; Kálinjar, ix.
McMorine, C., second siege of Bhurtpore, xiv.
McQuhae, W., Kamonah; Ganáori, ix. Nipál war, 2nd Division, x.
Montagu, E., first Mysore war, iii. Second Mysore war, iv. Note relative to his supersession of Colonel Smith, iv. App. E. Third Mysore war; is wounded at Seringapatam and dies, vi.
Montgomerie, T., Pindári war, xii.

Mordaunt, J., second Rohilla war; killed at Bitaurah, v.
Morland, R. S. B., Nipál war, 2nd Division, x. Háthras, xi. Pindári war, centre Division, xii. Second siege of Bhurtpore, xiv.
Morris, R., Máhrátá war; killed at Bela in Bandelkhand, vii.
Mullen, J., Matross, his gallant conduct at Dhálra, vii.

Nash, Sebastian, first Mysore war, iii. Second Mysore war; dies on service, iv.
Neish, J., first Mysore war, iii.
Nelly, J., first Mysore war, iii. Second Mysore war, iv. Máhrátá war; Aligarh, Delhi, Agra, Laswári, vii. Sieges of Deeg and Bhurtpore; takes a good shot; is wounded in the eye, viii.
Nicholl, T., second siege of Bhurtpore, xiv.

O'Hanlon, E. F., first Burmese war; killed at Kokkeing, xiii.
O'Laughlin, Sergeant, blows open the gate of Rámpura, vii.
Oliphant, W., Pindári war, xii. Second siege of Bhurtpore, xiv.

Palmer, C. H., China, vii.
Parker, C., second siege of Bhurtpore, xiv.
Parker, W., Máhrátá war; Agra, Laswári, vii. Sieges of Deeg and Bhurtpore, viii.
Parlby, S., Kamona; Ganáori; Java H.A., ix.
Paschaud, C. F., Bombay, vii.
Paschaud, J. F., Máhrátá war; battle of Deeg, viii.
Patch, C., Pindári war; dies at Betul, xii.
Paton, J., first Burmese war, xiii.
Pearse, T. D., first Mysore War; sent with a brigade of N.I. to Madras; deprived of the command by Sir E. Coote; Polilor; Sholingar; covers the march of the army to Vellore and back; goes to Bengal for his health; wounded at Cuddalore; honourably received by Warren Hastings on his return; presented with an honorary sword; orders on the occasion, iii.
Pennington, G., sen., third Mysore war, vi. First siege of Bhurtpore, viii. Nipál war, 2nd Division, x. Háthras, xi. Pindári war, xii.

Pennington, G., jun., Háthras, xi. Pindári war, xii.
· Pennington, Assistant-Surgeon, R. B., Pindári war, xii.
Percival, G., killed at the siege of Bhurtpore, viii.
Pereira, I., Mauritius, ix. Nipál war, 1st Division, x. Háthras, xi. Pindári war, xii. Second siege of Bhurtpore, xiv.
Perry, R., killed at or near Patna, i.
Pew, P. L., Nipál war, second campaign, x. Pindári war; Garhákota, Asirgárh, xii. Second siege of Bhurtpore, xiv.
Pillans, W. S., Barrackpore mutiny, xiii. Second siege of Bhurtpore, xiv.
Playfair, H. L., Nipál war, 2nd Division, x.
Pollock, G., Máhrátá war; battle and siege of Deeg; first siege of Bhurtpore, viii. Nipál war, 1st Division, x. First Burmese war, xiii.
Pryce, E., Kamonah; Ganáori; Bhawáni; Kálinjar, ix.

Raban (C.B.), G., Bombay detachment, ii. Máhrátá war; Aligarh, Delhi, Agra; detached with Colonel Don; Rámpura, vii. Sieges of Deeg and Bhurtpore, viii. Gohad, ix.
Ralfe, H., Java; Cornelis, ix.
Rattray, A., Bombay detachment, ii.
Rattray, W., Bombay detachment, ii.
Rawlins, J., Mauritius, ix. First Burmese war; Árákán, xiii.
Rawlinson, G. H., first Burmese war, xiii.
Read, J., killed at or near Patna, i.
Renny, Surgeon C., Pindári war, reserve, xii.
Revell, J. R., second siege of Bhurtpore, xiv.
Richards, W., third Mysore war, vi. Máhrátá war; Bandelkhand; Gwalior (wounded), vii. Java; Welter-vreeden; wounded at Cornelis, ix.
Roberts, Roderick, Nipál war; Colonel Gregory's detachment, x. Háthras, xi. Pindári War, xiii. Second siege of Bhurtpore, xiv.
Robinson, J., first Mysore war, iii. Sásni; Bijigarh; Kachaura; Máhrátá war; Aligarh, Delhi, Agra, vii.
Rodber, J., Nipál war, 2nd Division, x. Háthras, xi. Pindári war; remarkable march with 6th Troop H.A. to Siuni. Chándá, xii.

Rotton, J. S., second siege of Bhurtpore, xiv.

Sage, T. E., second siege of Bhurtpore, xiv.
Sampson, G. F. J., second Mysore war; wounded at Sátiya Mangalam; dies at Madras, iv.
Sanders, T., Háthras, xi. Pindári war; Dhámoni, Mandalah, Sátanwári, Garhákota, xii. Second siege of Bhurtpore, xiv.
Sands, R., operations near Benares; Bijigarh, ii.
Sconce, J., Nipál war, 2nd Division, x. Háthras, xi. Pindári war, xii.
Scott, C., Major Camac's detachment, ii.
Scott, G. R., Nipál war, 1st Division, x. Háthras, xi. Pindári war, xii.
Scott, J. W., first Burmese war; Rámu (wounded); Árákán, xiii.
Scott, Jonathan, Java; Weltervreeden; Cornelis, ix. First Burmese war; Assam, xiii. Second siege of Bhurtpore, xiv.
Sears, S., Bombay detachment, ii.
Shaw, S., Java; expeditions to Palimbang; Celebes, ix.
Shipton, W., operations near Benares; Bijigarh, ii. Second Mysore war, iv. Second Rohilla war, v. Sásni; Bijigarh; Kachaura; Máhrátá war, Aligarh (wounded), vii.
Sibbald, W., first Rohilla war, ii.
Silári, Lascar, Nipál war; his gallant conduct at Parsa, x.
Smith, C., Nipál war, 2nd Division, x. Háthras, xi. Pindári war; Táragarh, Mádhurájpura, Nasridah, xii. First Burmese war; Assam, xiii.
Smith (M.D.), Assistant-Surgeon G. G., Pindári war, left Division, xii.
Smith, James, second Mysore war; dies at Bangalore, iv.
Smith, John D., Máhrátá war; sieges of Deeg and Bhurtpore, viii. Java; Weltervreeden; Cornelis; dies on service, ix.
Sotheby, F. S., Pindári war; Mahidpore, xii.
Spilsbury, Assistant-Surgeon G. G., Pindári war, centre Division, xii.
Stark, H., Egypt, vi. Máhrátá war; joins Lake with H.A., vii. Pursuit of Holkar and battle of Fatehgarh;

sieges of Deeg and Bhurtpore, viii. Kamonah; Ganáori; Entáori, ix. Pindári war, xii. Second siege of Bhurtpore, xiv.

Swiney, G., Máhrátá war; battle and siege of Deeg; first siege of Bhurtpore (wounded), viii.

Syme, A., first Mysore war, iii.

Tennant, J., Kálinjar, ix. Nipál war, 3rd Division, x. Pindári war, centre Division, xii. Second siege of Bhurtpore, xiv.

Thompson, A., Pindári war, xii. First Burmese war; dies at Prome, xiii.

Tickell, R., Máhrátá war; battle and siege of Deeg; first siege of Bhurtpore, viii. Removed to the Engineers; Bhawáni; Entáori, ix. Nipál war, second campaign, x. Pindári war; Mandalah, Sátanwári, xii.

Tilfer, J., second Rohilla war; killed at Bitaurah, v.

Timbrell, (C.B.), T., Bandelkhand; wounded at Paririyah, ix. Nipál war, 3rd Division, x. Háthras, xi. Pindári war, centre Division, xii. First Burmese war; commands a flotilla on the Brahmaputra: sent to Rangoon; defence of the pagoda; returns home sick, xiii.

Timings, H., first Burmese war, xiii.

Todd, E. d'A., second siege of Bhurtpore, xiv.

Tollemach, W., Háthras, xi.

Tomkyns, J., first Mysore war, iii. Second Mysore war, iv. Third Mysore war, Seringapatam; Colonel Stevenson's force in Mysore, vi.

Toppin, J., second Mysore war, iv. Third Mysore war, vi.

Torckler, P. A., second siege of Bhurtpore, xiv.

Tulloh, R., second Mysore war, iv.

Turton, J., first Burmese war; Assam, xiii.

Turton, R., first Mysore war, iii. Gohad, ix.

Twemlow, G., Nipál war, 1st Division, x. Second campaign, x. Pindári war; Chándá, xii.

Vanrenen, T. A., Háthras, xi. Pindári war, xii.

Vernon, C., Major Popham's detachment, ii.

Wade, E. S. A. W. W., second siege of Bhurtpore, xiv.

Wakefield, J. W., second siege of Bhurtpore, xiv.

Walcott, W. G., Nipál war, second campaign (wounded), x. Pindári war, Nágpur subsidiary force, xii.
Walker, J., first Mysore war, iii.
Watkins, A., second Rohilla war, v.
Webb, H., Kálinjar, ix. Pindári war, xii.
Webb, N. S., Kálinjar, ix. Nipál war, 3rd Division, x. Barrackpore mutiny, xiii.
Whinfield, C. R., Háthras, xi. Pindári war, xii. Second siege of Bhurtpore, xiv.
Whish, W. S., Háthras, xi. Pindári war, xii. Second siege of Bhurtpore, xiv.
Wiggens, C. H., anecdote regarding, xiii. App. A. Second siege of Bhurtpore, xiv.
Wilkinson, J., first Mysore war, iii.
Wilson, A., second siege of Bhurtpore, xiv.
Wilson, R. B., Nipál war; Kamáon, x. Háthras, xi. Pindári war, xii.
Winbolt, W., second Mysore war; Sátiya Mangalam (wounded), iv. Ceylon, v. Máhrátá war; wounded at Shikoabád; detached with Colonel Monson's force, which retreats; is drowned in the Banás, vii.
Winwood, R., Giria; Udwah Nálá; Patna; Buxar, i.
Witherington, dies in the Black Hole, Calcutta, i.
Wittit, C., second Mysore war; refers question regarding precedence of Artillery, iv. Sent to Bombay; Máhrátá war, Bandelkhand; dies, vii.
Wood, H. J., Háthras, xi. Pindári war, xii. Second siege of Bhurtpore, xiv.
Woodburn, D., first Mysore war, iii. Second Mysore war, iv.
Woodrooffe, G. A., second siege of Bhurtpore, xiv.

INDEX TO NAMES OF MADRAS ARTILLERY OFFICERS.

Alcock, G., first Burmese war, xiii.
Aldritt, J., first Burmese war, xiii.

Begbie, P. J., first Burmese war, xiii.
Bell, J., second Mysore war, iv.
Bell, R., second Mysore war, iv. Third Mysore war, vi. Egypt, vi.
Bennett, J., Pindári war; Mahidpore, xii.
Bettson, J., second Mysore war, iv.
Black, C. W., Java; Jojokarta, ix.
Blundell, F., Pindári war, xii.
Bond, F., first Burmese war, xiii.
Bonner, J. G., Pindári war, xii.
Burgoyne, F., first Burmese war, xiii.
Burke, Ulick, second Mysore war, iv.
Burton, W. M., first Burmese war, xiii.

Carlisle, C., second Mysore war, iv.
Chisholm, W., killed at Korygám, xii.
Clarke, Tred., first Mysore war, iii. Second Mysore war, iv.
Cleaveland, S., Egypt, vi. Pindári war, xii.
Conan, W. N., killed at siege of Bangalore, iv.
Conran, G., Pindári war; Asirgárh, xii.
Cookesley, T., killed at Seringapatam, vi.
Coull, A. F., Pindári war; Nágpur, xii.
Coupland, W., second Mysore war, iv.
Court, M. H., Amboyna, ix.
Crosdill (C.B.), J., second Mysore war, iv. Pindári war; Nágpur, Asirgárh, xii.
Cullen, W., Bourbon, ix.
Cussans, T., Pindári war, xii.

Dalrymple, S., second Mysore war, iv. Pindári war, xii.
Daly, M., second Mysore war, iv.
Darke, R., second Mysore war, iv.
Dickenson, J., first Burmese war, xiii.
Donaldson, C., second Mysore war, iv.
Driffield, E. J. A., Java; dies of wounds at Weltervreeden, ix.

Fennell, Baker, second Mysore war, iv.
Freese, J. W., second Mysore war, iv.
Friend, Benj., second Mysore war, iv.
Frith, J. H., Pindári war; Ashti, Asirgárh, xii.

Gamage, J. J., Pindári war; Mahidpore, xii.
Geils, T., second Mysore war, iv.
Geils, T. E., first Burmese war, xiii.
Geoghegan, F., second Mysore war, iv.
Gibson, killed at the Bhor Ghát, ii.
Goreham, G. J., Pindári war; Nágpur; dies of exposure at the siege of Chándá, xii.
Gourlay, J., second Mysore war, iv.
Griffin, Troop Quarter-Master, Pindári war; Mahidpore, xii.

Hathway, J., second Mysore war, iv.
Hay, Gunner John, gallant conduct at siege of Bangalore, iv.
Hayes, T., first Mysore war, iii.
Hopkinson, C., first Burmese war, xiii.
Howley, R., second Mysore war, iv.
Hunter, N., Pindári war; Nágpur, Siuni, xii.

Isaacke, W. B., second Mysore war, iv.

Jourdan, J., third Mysore war; killed at Seringapatam, vi.

Kennan, T. Y. B., first Burmese war, xiii.
King, E., Pindári war; Nágpur, xii.

Lamb, J., first Burmese war; Árákán, xiii.
Lewis, W. F., Pindári war; Asirgárh, xii. First Burmese war; dies at Prome, xiii.

Ley, J. M., Pindári war; Nágpur, xii.
Limond, Sir J., Java, ix.

Macintire, A., second Mysore war, iv.
Macintosh, B., Pindári war, xii.
Mackay, A., second Mysore war, iv.
Mackay, D., first Mysore war; Cuddalore, iii.
Mackie, J., second Mysore war, iv.
Mandeville, F., second Mysore war, iv.
Maxwell, J., Pindári war; Sitabaldi, Nágpur, xii.
Middlecoat, G., first Burmese war; Árákán, xiii.
Montgomerie, P., Pindári war; Nágpur, Chándá, xii. First Burmese war, xiii.
Moore, R. C., first Burmese war, xiii.
Moore, Teerad, second Mysore war, iv.
Moorhouse, J., first Mysore war; blows open a gate at Chillambram; assists in recapturing a gun, iii. Second Mysore war; killed at Bangalore; his character, iv.
Morison, W., Pindári war, xii.
Morris, A., second Mysore war, iv.
Munro, E. S., Java; wounded at Cornelis, ix.
Murray, A. L., first Burmese war, xiii.

Neilson, J., second Mysore war, iv.
Noble (C.B.), J., Java; Weltervreeden, Cornelis, ix. Pindári war; Mahidpore, xii.
Noble, T. G., Pindári war; Mahidpore, xii.

O'Brien, F., second Mysore war, iv.
O'Brien, Matross, gallant conduct of, ix.
Onslow, G. W., first Burmese war, xiii.

Patterson, J. C., first Burmese war, xiii.
Penny, Surgeon, killed at Bhor Ghát, ii.
Poggenpohl, P., Pindári war; Nágpur, xii.
Poignand, C. W., Pindári war; Nágpur, Asirgárh, xii.
Prescott, F., second Mysore war, iv. Third Mysore war, vi.

Rudyerd, H. T., Java; Jojokarta, ix. Pindári war; Mahidpore, xii.

Russell, J., second Mysore war, iv.
Rutledge, F. W., killed at the Bhor Ghát, ii.

Saxon, G., second Mysore war, iv.
Scott, J. G., Egypt, vi.
Seton, R. S., first Burmese war, xiii.
Sheriff, Æ., Pindári war, xii.
Slipper, J., first Mysore war; second Mysore war; killed at Bangalore, iv.
Smith, D., first Mysore war; made prisoner, iii. Second Mysore war, iv. Third Mysore war, vi.
Speediman, R., first Mysore war; joins Colonel Pearse, iii. Second Mysore war, iv.
Symes, G. F., first Burmese war, xiii.

Tanner, J. A., second Mysore war, iv.
Taynton, J., second Mysore war, iv.

Weldon, A., Pindári war; Nágpur, Asirgárh, xii.
Whinyates, F. F., Pindári war; Asirgárh, xii.

INDEX TO
NAMES OF BOMBAY ARTILLERY OFFICERS.

Bellasis, C. B., Pindári war, xii.
Bond, C. J., Pindári war, xi.

Griffith, J. G., Pindári war, xii.

Hardy, E., Pindári war, xii.
Hessman, H., Pindári war, xii.

Jacob, W., Pindári war, xii.
Jones (K.C.B.), Sir R., second Mysore war; commands Bombay Artillery, iv. Máhrátá war; joins Lake at Bhurtpore with a force from Bombay, vii.

Lawnan, G. A., commands Bombay Artillery in third Mysore war, vi.
Lawrie, J., Pindári war, xii.
Lyons, G. R., Pindári war, xii.

McRedie, W., killed at Seringapatam, vi.
Miller, W., Pindári war, xii.

Osborne, H. L., Pindári war, xii.

Pierce, F. H., Pindári war, xii.
Powell, G., Egypt, iv.

Russell Lechmere, C., first Burmese war, xiii.

Stevenson, T., Pindári war, xii.
Strover, S. R., Pindári war, iv.

Thew, R., Pindári war, xii.
Torriano, A., third Mysore war; killed at Seringapatam, vi.
Torriano, J. S., joins Goddard's force at Surat, ii.

Warden, G., Egypt, vi.

INDEX TO
NAMES OF ROYAL ARTILLERY OFFICERS.

Barker, Sir R., comes to Bengal; goes to Madras, i. Note relative to, i. App. A.
Beevor, R., Egypt, vi.
Byers, J. S., Java; Cornelis, Jojokarta, ix. Nipâl war, 2nd Division, x.

Clarke, R., second Mysore war; Seringapatam, iv.
Colebrooke, W. M. G., Java; Weltervreeden, Cornelis, ix.

Desbrisay, T., Ceylon, v.

Gold, R., second Mysore war, iv.

Hamilton, G. L., second Mysore war, iv.
Hunter, R., second Mysore war, iv.

Napier, C. F., Java; Weltervreeden, Cornelis, ix.
Nicolay, W., second Mysore war; transferred to Royal Engineers, iv.

Patton, P., Java; Weltervreeden; killed at Cornelis, ix.

Ross, T., second Mysore war, iv.

Scott, D., second Mysore war; dies at Seringapatam, iv.

Terrott, C., second Mysore war, iv.

Worsley, R. V., Ceylon, v.

PRINTED AT THE CAXTON PRESS, BECCLES.

65, *Cornhill, and* 1, *Paternoster Square, London,*
October, 1876.

A LIST OF HENRY S. KING AND CO.'S PUBLICATIONS.

ABBEY (Henry).

Ballads of Good Deeds, and Other Verses. Fcap. 8vo. Cloth gilt, price 5*s*.

ABDULLA (Hakayit).

Autobiography of a Malay Munshi. Translated by J. T. Thomson, F.R.G.S. With Photo-lithograph Page of Abdulla's Manuscript. Post 8vo. Cloth, price 12*s*.

ADAMS (A. L.), M.A., M.B, F.R.S., F.G.S.

Field and Forest Rambles of a Naturalist in New Brunswick. With Notes and Observations on the Natural History of Eastern Canada. Illustrated. 8vo. Cloth, price 14*s*.

ADAMS (F. O.), F.R.G.S.

The History of Japan. From the Earliest Period to the Present Time. New Edition, revised. 2 volumes. With Maps and Plans. Demy 8vo. Cloth, price 21*s*. each.

ADAMS (W. D., Jun.).

Lyrics of Love, from Shakespeare to Tennyson. Selected and arranged by. Fcap. 8vo. Cloth extra, gilt edges, price 3*s*. 6*d*.

ADAMS (John), M.A.

St. Malo's Quest, and other Poems. Fcap. 8vo. Cloth, 5*s*.

ADON.

Through Storm & Sunshine. Illustrated by M. E. Edwards, A. T. H. Paterson, and the Author. Crown 8vo. Cloth, price 7*s*. 6*d*.

A. K. H. B.

A Scotch Communion Sunday, to which are added Certain Discourses from a University City. By the Author of "The Recreations of a Country Parson." Second Edition. Crown 8vo. Cloth, price 5*s*.

ALLEN (Rev. R.), M.A.

Abraham; his Life, Times, and Travels, as told by a Contemporary 3,800 years ago. With Map. Post 8vo. Cloth, price 10*s*. 6*d*.

AMOS (Prof. Sheldon).

Science of Law. Second Edition. Crown 8vo. Cloth, price 5*s*.

Volume X. of The International Scientific Series.

ANDERSON (Rev. C.), M.A.

New Readings of Old Parables. Demy 8vo. Cloth, price 4*s*. 6*d*.

Church Thought and Church Work. Edited by. Second Edition. Demy 8vo. Cloth, price 7*s*. 6*d*.

Words and Works in a London Parish. Edited by. Second Edition. Demy 8vo. Cloth, price 6*s*.

The Curate of Shyre. Second Edition. 8vo. Cloth, price 7*s*. 6*d*.

ANDERSON (Col. R. P.).

Victories and Defeats. An Attempt to explain the Causes which have led to them. An Officer's Manual. Demy 8vo. Cloth, price 14*s*.

ANDERSON (R. C.), C.E.

Tables for Facilitating the Calculation of every Detail in connection with Earthen and Masonry Dams. Royal 8vo. Cloth, price £2 2*s*.

ANSON (Lieut.-Col. The Hon. A.), V.C., M.P.

The Abolition of Purchase and the Army Regulation Bill of 1871. Crown 8vo. Sewed, price 1*s*.

Army Reserves and Militia Reforms. Crown 8vo. Sewed, price 1*s*.

Story of the Supersessions. Crown 8vo. Sewed, price 6*d*.

ARCHER (Thomas).

About my Father's Business. Work amidst the Sick, the Sad, and the Sorrowing. Crown 8vo. Cloth, price 5*s*.

ARGYLE (Duke of).

Speeches on the Second Reading of the Church Patronage (Scotland) Bill in the House of Lords, June 2, 1874; and Earl of Camperdown's Amendment, June 9, 1874, placing the Election of Ministers in the hands of Ratepayers. Crown 8vo. Sewed, price 1*s*.

Army of the North German Confederation.

A Brief Description of its Organization, of the Different Branches of the Service and their *rôle* in War, of its Mode of Fighting, &c., &c. Translated from the Corrected Edition, by permission of the Author, by Colonel Edward Newdigate. Demy 8vo. Cloth, price 5*s*.

Ashantee War (The).

A Popular Narrative. By the Special Correspondent of the "Daily News." Crown 8vo. Cloth, price 6*s*.

ASHTON (J.).

Rough Notes of a Visit to Belgium, Sedan, and Paris, in September, 1870-71. Crown 8vo. Cloth, price 3*s*. 6*d*.

Aunt Mary's Bran Pie.

By the author of "St. Olave's." Illustrated. Cloth, price 3*s*. 6*d*.

Aurora.

A Volume of Verse. Fcap. 8vo. Cloth, price 5*s*.

AYRTON (J. C.).

A Scotch Wooing. 2 vols. Crown 8vo. Cloth.

BAGEHOT (Walter).

Physics and Politics; or, Thoughts on the Application of the Principles of "Natural Selection" and "Inheritance" to Political Society. Third Edition. Crown 8vo. Cloth, price 4*s*.

Volume II. of The International Scientific Series.

The English Constitution. A New Edition, Revised and Corrected, with an Introductory Dissertation on Recent Changes and Events. Crown 8vo. Cloth, price 7*s*. 6*d*.

Lombard Street. A Description of the Money Market. Sixth Edition. Crown 8vo. Cloth, price 7*s*. 6*d*.

BAIN (Alexander), LL.D.

Mind and Body: the Theories of their relation. Fifth Edition. Crown 8vo. Cloth, price 4*s*.

Volume IV. of The International Scientific Series.

BANKS (Mrs. G. L.).

God's Providence House. New Edition. Crown 8vo. Cloth, price 3*s*. 6*d*.

BARING (T. C.), M.A., M.P.

Pindar in English Rhyme. Being an Attempt to render the Epinikian Odes with the principal remaining Fragments of Pindar into English Rhymed Verse. Small Quarto. Cloth, price 7*s*.

BARLEE (Ellen).

Locked Out: a Tale of the Strike. With a Frontispiece. Royal 16mo. Cloth, price 1*s*. 6*d*.

BAUR (Ferdinand), Dr. Ph., Professor in Maulbronn.

A Philological Introduction to Greek and Latin for Students. Translated and adapted from the German of. By C. Kegan Paul, M.A. Oxon.; and the Rev. E. D. Stone, M.A., late Fellow of King's College, Cambridge, and Assistant Master at Eton. Crown 8vo. Cloth, price 6*s*.

BAYNES (Rev. Canon R. H.), M.A.

Home Songs for Quiet Hours. Third Edition. Fcap. 8vo. Cloth extra, price 3*s.* 6*d.*
This may also be had handsomely bound in Morocco with gilt edges.

BECKER (Bernard H.).

The Scientific Societies of London. Crown 8vo. Cloth, price 5*s.*

BENNETT (Dr. W. C.).

Baby May. Home Poems and Ballads. With Frontispiece. Crown 8vo. Cloth elegant, price 6*s.*

Baby May and Home Poems. Fcap. 8vo. Sewed in Coloured Wrapper, price 1*s.*

Narrative Poems & Ballads. Fcap. 8vo. Sewed in Coloured Wrapper, price 1*s.*

Songs for Sailors. Dedicated by Special Request to H. R. H. the Duke of Edinburgh. With Steel Portrait and Illustrations. Crown 8vo. Cloth, price 3*s.* 6*d.*
An Edition in Illustrated Paper Covers, price 1*s.*

Songs of a Song Writer. Crown 8vo. Cloth, price 6*s.*

BENNIE (Rev. J. N.), M.A.

The Eternal Life. Sermons preached during the last twelve years. Crown 8vo. Cloth, price 6*s.*

BERNARD (Bayle).

Samuel Lover, the Life and Unpublished Works of. In 2 vols. With a Steel Portrait. Post 8vo. Cloth, price 21*s.*

BERNSTEIN (Prof.).

The Five Senses of Man. With 91 Illustrations. Second Edition. Crown 8vo. Cloth, price 5*s.*
Volume XXI. of The International Scientific Series.

BETHAM - EDWARDS (Miss M.).

Kitty. With a Frontispiece. Crown 8vo. Cloth, price 3*s.* 6*d.*

Mademoiselle Josephine's Fridays, and Other Stories. Crown 8vo. Cloth, price 7*s.* 6*d.*

BISCOE (A. C.).

The Earls of Middleton, Lords of Clermont and of Fettercairn, and the Middleton Family. Crown 8vo. Cloth, price 10*s.* 6*d.*

BLANC (H.), M.D.

Cholera: How to Avoid and Treat it. Popular and Practical Notes. Crown 8vo. Cloth, price 4*s.* 6*d.*

BLASERNA (Prof. Pietro).

The Theory of Sound in its Relation to Music. With numerous Illustrations. Crown 8vo. Cloth, price 5*s.*
Volume XXII. of The International Scientific Series.

BLUME (Major W.).

The Operations of the German Armies in France, from Sedan to the end of the war of 1870-71. With Map. From the Journals of the Head-quarters Staff. Translated by the late E. M. Jones, Maj. 20th Foot, Prof. of Mil. Hist., Sandhurst. Demy 8vo. Cloth, price 9*s.*

BOGUSLAWSKI (Capt. A. von).

Tactical Deductions from the War of 1870-71. Translated by Colonel Sir Lumley Graham, Bart., late 18th (Royal Irish) Regiment. Third Edition, Revised and Corrected. Demy 8vo. Cloth, price 7*s.*

BONWICK (J.), F.R.G.S.

The Tasmanian Lily. With Frontispiece. Crown 8vo. Cloth, price 5*s.*

Mike Howe, the Bushranger of Van Diemen's Land. With Frontispiece. Crown 8vo. Cloth price 5*s.*

BOSWELL (R. B.), M.A., Oxon.

Metrical Translations from the Greek and Latin Poets, and other Poems. Crown 8vo. Cloth, price 5*s*.

BOTHMER (Countess von).

Cruel as the Grave. A Novel. 3 vols. Crown 8vo. Cloth.

BOWEN (H. C.), M.A., Head Master of the Grocers' Company's Middle Class School at Hackney.

Studies in English, for the use of Modern Schools. Small Crown 8vo. Cloth, price 1*s*. 6*d*.

BOWRING (L.), C.S.I.

Eastern Experiences. Illustrated with Maps and Diagrams. Demy 8vo. Cloth, price 16*s*.

BRADLEY (F. H.).

Ethical Studies. Critical Essays in Moral Philosophy. Large post 8vo. Cloth, price 9*s*.

Brave Men's Footsteps. By the Editor of "Men who have Risen." A Book of Example and Anecdote for Young People. With Four Illustrations by C. Doyle. Third Edition. Crown 8vo. Cloth, price 3*s*. 6*d*.

BRIALMONT (Col. A.).

Hasty Intrenchments. Translated by Lieut. Charles A. Empson, R.A. With Nine Plates. Demy 8vo. Cloth, price 6*s*.

Briefs and Papers. Being Sketches of the Bar and the Press. By Two Idle Apprentices. Second Edition. Crown 8vo. Cloth, price 7*s*. 6*d*.

BROOKE (Rev. J. M. S.), M. A.

Heart, be Still. A Sermon preached in Holy Trinity Church, Southall. Imperial 32mo. Sewed, price 6*d*.

BROOKE (Rev. S. A.), M. A., Chaplain in Ordinary to Her Majesty the Queen, and Minister of Bedford Chapel, Bloomsbury.

The Late Rev. F. W. Robertson, M.A., Life and Letters of. Edited by.

I. Uniform with the Sermons. 2 vols. With Steel Portrait. Price 7*s*. 6*d*.

II. Library Edition. 8vo. With Two Steel Portraits. Price 12*s*.

III. A Popular Edition, in 1 vol. 8vo. Price 6*s*.

Theology in the English Poets. — COWPER, COLERIDGE, WORDSWORTH, and BURNS. Third Edition. Post 8vo. Cloth, price 9*s*.

Christ in Modern Life. Ninth Edition. Crown 8vo. Cloth, price 7*s*. 6*d*.

Sermons. First Series. Ninth Edition. Crown 8vo. Cloth, price 6*s*.

Sermons. Second Series. Third Edition. Crown 8vo. Cloth, price 7*s*.

Frederick Denison Maurice: The Life and Work of. A Memorial Sermon. Crown 8vo. Sewed, price 1*s*.

BROOKE (W. G.), M.A.

The Public Worship Regulation Act. With a Classified Statement of its Provisions, Notes, and Index. Third Edition, revised and corrected. Crown 8vo. Cloth, price 3*s*. 6*d*.

Six Privy Council Judgments—1850-1872. Annotated by. Third Edition. Crown 8vo. Cloth, price 9*s*.

BROUN (J. A.).

Magnetic Observations at Trevandrum and Augustia Malley. Vol. I. 4to. Cloth, price 63*s*.

The Report from above, separately sewed, price 21*s*.

BROWN (Rev. J. Baldwin), B.A.

The Higher Life. Its Reality, Experience, and Destiny. Fourth Edition. Crown 8vo. Cloth, price 7s. 6d.

Doctrine of Annihilation in the Light of the Gospel of Love. Five Discourses. Second Edition. Crown 8vo. Cloth, price 2s. 6d.

BROWN (J. Croumbie), LL.D.

Reboisement in France; or, Records of the Replanting of the Alps, the Cevennes, and the Pyrenees with Trees, Herbage, and Bush. Demy 8vo. Cloth, price 12s. 6d.

The Hydrology of Southern Africa. Demy 8vo. Cloth, price 10s. 6d.

BROWNE (Rev. M. E.)

Until the Day Dawn. Four Advent Lectures. Crown 8vo. Cloth, price 2s. 6d.

BRYANT (W. C.)

Poems. Red-line Edition. With 24 Illustrations and Portrait of the Author. Crown 8vo. Cloth extra, price 7s. 6d.

A Cheaper Edition, with Frontispiece. Small crown 8vo. Cloth, price 3s. 6d.

BUCHANAN (Robert).

Poetical Works. Collected Edition, in 3 vols., with Portrait. Crown 8vo. Cloth, price 6s. each.

Master-Spirits. Post 8vo. Cloth, price 10s. 6d.

BULKELEY (Rev. H. J.).

Walled in, and other Poems. Crown 8vo. Cloth, price 5s.

BUNNETT (F. E.).

Linked at Last. Crown 8vo. Cloth.

BURTON (Mrs. Richard).

The Inner Life of Syria, Palestine, and the Holy Land. With Maps, Photographs, and Coloured Plates. 2 vols. Second Edition. Demy 8vo. Cloth, price 24s.

CADELL (Mrs. H. M.).

Ida Craven: A Novel. 2 vols. Crown 8vo. Cloth.

CALDERON.

Calderon's Dramas: The Wonder-Working Magician,—Life is a Dream—The Purgatory of St. Patrick. Translated by Denis Florence MacCarthy. Post 8vo. Cloth, price 10s.

CARLISLE (A. D.), B. A.

Round the World in 1870. A Volume of Travels, with Maps. New and Cheaper Edition. Demy 8vo. Cloth, price 6s.

CARNE (Miss E. T.).

The Realm of Truth. Crown 8vo. Cloth, price 5s. 6d.

CARPENTER (E.).

Narcissus and other Poems. Fcap. 8vo. Cloth, price 5s.

CARPENTER (W. B.), LL.D., M.D., F.R.S., &c.

The Principles of Mental Physiology. With their Applications to the Training and Discipline of the Mind, and the Study of its Morbid Conditions. Illustrated. Fourth Edition. 8vo. Cloth, price 12s.

CARR (Lisle).

Judith Gwynne. 3 vols. Second Edition. Crown 8vo. Cloth.

CHRISTOPHERSON (The late Rev. Henry), M.A.

Sermons. With an Introduction by John Rae, LL.D., F.S.A. First Series. Crown 8vo. Cloth, price 7s. 6d.

Sermons. With an Introduction by John Rae, LL.D., F.S.A. Second Series. Crown 8vo. Cloth, price 6s.

CLAYTON (Cecil).

Effie's Game; How She Lost and How She Won. A Novel. 2 vols. Cloth.

CLERK (Mrs. Godfrey).

'Ilâm en Nâs. Historical Tales and Anecdotes of the Times of the Early Khalifahs. Translated from the Arabic Originals. Illustrated with Historical and Explanatory Notes. Crown 8vo. Cloth, price 7*s.*

CLERY (C.), Capt.

Minor Tactics. With 26 Maps and Plans. Second Edition. Demy 8vo. Cloth, price 16*s.*

CLODD (Edward), F.R.A.S.

The Childhood of the World: a Simple Account of Man in Early Times. Third Edition. Crown 8vo. Cloth, price 3*s.*

A Special Edition for Schools. Price 1*s.*

The Childhood of Religions. Including a Simple Account of the Birth and Growth of Myths and Legends. Crown 8vo. Cloth, price 5*s.*

COLERIDGE (Sara).

Pretty Lessons in Verse for Good Children, with some Lessons in Latin, in Easy Rhyme. A New Edition. Illustrated. Fcap. 8vo. Cloth, price 3*s.* 6*d.*

Phantasmion. A Fairy Tale. With an Introductory Preface by the Right Hon. Lord Coleridge, of Ottery St. Mary. A New Edition. Illustrated. Crown 8vo. Cloth, price 7*s.* 6*d.*

Memoir and Letters of Sara Coleridge. Edited by her Daughter. With Index. 2 vols. With Two Portraits. Third Edition, Revised and Corrected. Crown 8vo. Cloth, price 24*s.*

Cheap Edition. With one Portrait. Cloth, price 7*s.* 6*d.*

COLLINS (Mortimer).

The Princess Clarice. A Story of 1871. 2 vols. Cloth.

Squire Silchester's Whim. 3 vols. Cloth.

Miranda. A Midsummer Madness. 3 vols. Cloth.

Inn of Strange Meetings, and other Poems. Crown 8vo. Cloth, price 5*s.*

The Secret of Long Life. Dedicated by special permission to Lord St. Leonards. Fourth Edition. Large crown 8vo. Cloth, price 5*s.*

COLLINS (Rev. R.), M.A.

Missionary Enterprise in the East. With special reference to the Syrian Christians of Malabar, and the results of modern Missions. With Four Illustrations. Crown 8vo. Cloth, price 6*s.*

CONGREVE (Richard), M.A., M.R.C.P.L.

Human Catholicism. Two Sermons delivered at the Positivist School on the Festival of Humanity, 87 and 88, January 1, 1875 and 1876. Demy 8vo. Sewed, price 1*s.*

CONWAY (Moncure D.).

Republican Superstitions. Illustrated by the Political History of the United States. Including a Correspondence with M. Louis Blanc. Crown 8vo. Cloth, price 5*s.*

CONYERS (Ansley).

Chesterleigh. 3 vols. Crown 8vo. Cloth.

COOKE (M. C.), M.A., LL.D.

Fungi; their Nature, Influences, Uses, &c. Edited by the Rev. M. J. Berkeley, M.A., F.L.S. With Illustrations. Second Edition. Crown 8vo. Cloth, price 5*s.*

Volume XIV. of The International Scientific Series.

COOKE (Prof. J. P.), of the Harvard University.

The New Chemistry. With 31 Illustrations. Third Edition. Crown 8vo. Cloth, price 5*s*.
Volume IX. of The International Scientific Series.

Scientific Culture. Crown 8vo. Cloth, price 1*s*.

COOPER (T. T.), F.R.G.S.

The Mishmee Hills: an Account of a Journey made in an Attempt to Penetrate Thibet from Assam, to open New Routes for Commerce. Second Edition. With Four Illustrations and Map. Post 8vo. Cloth, price 10*s*. 6*d*.

Cornhill Library of Fiction (The). Crown 8vo. Cloth, price 3*s*. 6*d*. per volume.

Half-a-Dozen Daughters. By J. Masterman.
The House of Raby. By Mrs. G. Hooper.
A Fight for Life. By Moy Thomas.
Robin Gray. By Charles Gibbon.
Kitty. By Miss M. Betham-Edwards.
One of Two; or, The Left-Handed Bride. By J. Hain Friswell.
Ready - Money Mortiboy. A Matter-of-Fact Story.
God's Providence House. By Mrs. G. L. Banks.
For Lack of Gold. By Charles Gibbon.
Abel Drake's Wife. By John Saunders.
Hirell. By John Saunders.

CORY (Lieut. Col. Arthur).

The Eastern Menace; or, Shadows of Coming Events. Crown 8vo. Cloth, price 5*s*.

Cosmos.

A Poem. Fcap. 8vo. Cloth, price 3*s*. 6*d*.

COTTON (R. T.).

Mr. Carington. A Tale of Love and Conspiracy. 3 vols. Crown 8vo. Cloth.

CRESSWELL (Mrs. G.).

The King's Banner. Drama in Four Acts. Five Illustrations. 4to. Cloth, price 10*s*. 6*d*.

CROMPTON (Henry).

Industrial Conciliation. Fcap. 8vo. Cloth, price 2*s*. 6*d*.

CUMMINS (H. I.), M. A.

Parochial Charities of the City of London. Sewed, price 1*s*.

CURWEN (Henry).

Sorrow and Song: Studies of Literary Struggle. Henry Mürger—Novalis—Alexander Petőfi—Honoré de Balzac—Edgar Allan Poe—André Chénier. 2 vols. Crown 8vo. Cloth, price 15*s*.

DANCE (Rev. C. D.).

Recollections of Four Years in Venezuela. With Three Illustrations and a Map. Crown 8vo. Cloth, price 7*s*. 6*d*.

D'ANVERS (N. R.).

The Suez Canal: Letters and Documents descriptive of its Rise and Progress in 1854-56. By Ferdinand de Lesseps. Translated by. Demy 8vo. Cloth, price 10*s*. 6*d*.

Little Minnie's Troubles. An Every-day Chronicle. With Four Illustrations by W. H. Hughes. Fcap. Cloth, price 3*s*. 6*d*.

DAVIDSON (Rev. Samuel), D.D., LL.D.

The New Testament, translated from the Latest Greek Text of Tischendorf. A new and thoroughly revised Edition. Post 8vo. Cloth, price 10*s*. 6*d*.

Canon of the Bible: Its Formation, History, and Fluctuations. Small crown 8vo. Cloth, price 5*s*.

DAVIES (G. Christopher).

Mountain, Meadow, and Mere: a Series of Outdoor Sketches of Sport, Scenery, Adventures, and Natural History. With Sixteen Illustrations by Bosworth W. Harcourt. Crown 8vo. Cloth, price 6*s*.

Rambles and Adventures of Our School Field Club. With Four Illustrations. Crown 8vo. Cloth, price 5*s*.

DAVIES (Rev. J. L.), M.A.

Theology and Morality. Essays on Questions of Belief and Practice. Crown 8vo. Cloth, price 7*s*. 6*d*.

DE KERKADEC (Vicomtesse Solange).

A Chequered Life, being Memoirs of the Vicomtesse de Leoville Meilhan. Edited by. Crown 8vo. Cloth, price 7*s*. 6*d*.

DE L'HOSTE (Col. E. P.).

The Desert Pastor, Jean Jarousseau. Translated from the French of Eugène Pelletan. With a Frontispiece. New Edition. Fcap. 8vo. Cloth, price 3*s*. 6*d*.

DE REDCLIFFE (Viscount Stratford), P.C., K.G., G.C.B.

Why am I a Christian? Fifth Edition. Crown 8vo. Cloth, price 3*s*.

DE TOCQUEVILLE (A.).

Correspondence and Conversations of, with Nassau William Senior, from 1834 to 1859. Edited by M. C. M. Simpson. 2 vols. Post 8vo. Cloth, price 21*s*.

DE VERE (Aubrey).

Alexander the Great. A Dramatic Poem. Small crown 8vo. Cloth, price 5*s*.

The Infant Bridal, and Other Poems. A New and Enlarged Edition. Fcap. 8vo. Cloth, price 7*s*. 6*d*.

DE VERE (Aubrey)—*continued:*

The Legends of St. Patrick, and Other Poems. Small crown 8vo. Cloth, price 5*s*.

St. Thomas of Canterbury. A Dramatic Poem. Large fcap. 8vo. Cloth, price 5*s*.

DE WILLE (E.).

Under a Cloud; or, Johannes Olaf. A Novel. Translated by F. E. Bunnètt. 3 vols. Crown 8vo. Cloth.

DENNIS (J.).

English Sonnets. Collected and Arranged. Elegantly bound. Fcap. 8vo. Cloth, price 3*s*. 6*d*.

DOBSON (Austin).

Vignettes in Rhyme and Vers de Société. Second Edition. Fcap. 8vo. Cloth, price 5*s*.

DONNÉ (A.), M.D.

Change of Air and Scene. A Physician's Hints about Doctors, Patients, Hygiene, and Society; with Notes of Excursions for Health. Second Edition. Large post 8vo. Cloth, price 9*s*.

DOWDEN (Edward), LL.D.

Shakspere: a Critical Study of his Mind and Art. Second Edition. Post 8vo. Cloth, price 12*s*.

Poems. Fcap. 8vo. Cloth, price 5*s*.

DOWNTON (Rev. H.), M.A.

Hymns and Verses. Original and Translated. Small crown 8vo. Cloth, price 3*s*. 6*d*.

DRAPER (J. W.), M.D., LL.D., Professor in the University of New York.

History of the Conflict between Religion and Science. Seventh Edition. Crown 8vo. Cloth, price 5*s*.

Volume XIII. of The International Scientific Series.

DREW (Rev. G. S.), M.A.

Scripture Lands in connection with their History. Second Edition. 8vo. Cloth, price 10*s.* 6*d.*

Nazareth: Its Life and Lessons. Third Edition. Crown 8vo. Cloth, price 5*s.*

The Divine Kingdom on Earth as it is in Heaven. 8vo. Cloth, price 10*s.* 6*d.*

The Son of Man: His Life and Ministry. Crown 8vo. Cloth, price 7*s.* 6*d.*

DREWRY (G. O.), M.D.

The Common-Sense Management of the Stomach. Third Edition. Fcap. 8vo. Cloth, price 2*s.* 6*d.*

DREWRY (G. O.), M.D., and BARTLETT (H. C.), Ph.D., F.C.S.

Cup and Platter: or, Notes on Food and its Effects. Small 8vo. Cloth, price 2*s.* 6*d.*

DURAND (Lady).

Imitations from the German of Spitta and Terstegen. Fcap. 8vo. Cloth, price 4*s.*

DU VERNOIS (Col. von Verdy).

Studies in leading Troops. An authorized and accurate Translation by Lieutenant H. J. T. Hildyard, 71st Foot. Parts I. and II. Demy 8vo. Cloth, price 7*s.*

EDEN (Frederick).

The Nile without a Dragoman. Second Edition. Crown 8vo. Cloth, price 7*s.* 6*d.*

EDWARDS (Rev. Basil).

Minor Chords; Or, Songs for the Suffering: a Volume of Verse. Fcap. 8vo. Cloth, price 3*s.* 6*d.*; paper, price 2*s.* 6*d.*

EILOART (Mrs.).

Lady Moretoun's Daughter. vols. Crown 8vo. Cloth.

ELLIOTT (Ebenezer), The Corn Law Rhymer.

Poems. Edited by his son, the Rev. Edwin Elliott, of St. John's, Antigua. 2 vols. Crown 8vo. Cloth, price 18*s.*

ENGLISH CLERGYMAN.

An Essay on the Rule of Faith and Creed of Athanasius. Shall the Rubric preceding the Creed be removed from the Prayer-book? Sewed. 8vo. Price 1*s.*

Epic of Hades (The).

By a New Writer. Author of "Songs of Two Worlds." Fcap. 8vo. Cloth, price 5*s.*

Eros Agonistes.

Poems. By E. B. D. Fcap. 8vo. Cloth, price 3*s.* 6*d.*

Essays on the Endowment of Research.

By Various Writers.

List of Contributors.

Mark Pattison, B. D.
James S. Cotton, B. A.
Charles E. Appleton, D. C. L.
Archibald H. Sayce, M. A.
Henry Clifton Sorby, F. R. S.
Thomas K. Cheyne, M. A.
W. T. Thiselton Dyer, M. A.
Henry Nettleship, M. A.

Square crown octavo. Cloth, price 10*s.* 6*d.*

EVANS (Mark).

The Story of our Father's Love, told to Children; being a New and Enlarged Edition of Theology for Children. With Four Illustrations. Fcap. 8vo. Cloth, price 3*s.* 6*d.*

A Book of Common Prayer and Worship for Household Use, compiled exclusively from the Holy Scriptures. Fcap. 8vo. Cloth, price 2*s.* 6*d.*

EYRE (Maj.-Gen. Sir V.), C.B., K.C.S.I., &c.

Lays of a Knight-Errant in many Lands. Square crown 8vo. With Six Illustrations. Cloth, price 7*s.* 6*d.*

FAITHFULL (Mrs. Francis G.).

Love Me, or Love Me Not. 3 vols. Crown 8vo. Cloth.

FARQUHARSON (M.).

I. Elsie Dinsmore. Crown 8vo. Cloth, price 3*s.* 6*d.*

II. Elsie's Girlhood. Crown 8vo. Cloth, price 3*s.* 6*d.*

III. Elsie's Holidays at Roselands. Crown 8vo. Cloth, price 3*s.* 6*d.*

FAVRE (Mons. J.).

The Government of the National Defence. From the 30th June to the 31st October, 1870. Translated by H. Clark. Demy 8vo. Cloth, price 10*s.* 6*d.*

FERRIS (Henry Weybridge).

Poems. Fcap. 8vo. Cloth, price 5*s.*

FISHER (Alice).

His Queen. 3 vols. Crown 8vo. Cloth.

FOOTMAN (Rev. H.), M.A.

From Home and Back; or, Some Aspects of Sin as seen in the Light of the Parable of the Prodigal. Crown 8vo. Cloth, price 5*s.*

FORBES (A.).

Soldiering and Scribbling. A Series of Sketches. Crown 8vo. Cloth, price 7*s.* 6*d.*

FOTHERGILL (Jessie).

Healey. A Romance. 3 vols. Crown 8vo. Cloth.

FOWLE (Rev. T. W.), M.A.

The Reconciliation of Religion and Science. Being Essays on Immortality, Inspiration, Miracles, and the Being of Christ. Demy 8vo. Cloth, price 10*s.* 6*d.*

FOX-BOURNE (H. R.).

The Life of John Locke, 1632—1704. 2 vols. Demy 8vo. Cloth, price 28*s.*

FRASER (Donald).

Exchange Tables of Sterling and Indian Rupee Currency, upon a new and extended system, embracing Values from One Farthing to One Hundred Thousand Pounds, and at Rates progressing, in Sixteenths of a Penny, from 1*s.* 9*d.* to 2*s.* 3*d.* per Rupee. Royal 8vo. Cloth, price 10*s.* 6*d.*

FRERE (Sir H. Bartle E.), G.C.B., G.C.S.I.

The Threatened Famine in Bengal: How it may be Met, and the Recurrence of Famines in India Prevented. Being No. 1 of "Occasional Notes on Indian Affairs." With 3 Maps. Crown 8vo. Cloth, price 5*s.*

FRISWELL (J. Hain).

The Better Self. Essays for Home Life. Crown 8vo. Cloth, price 6*s.*

One of Two; or, The Left-Handed Bride. With a Frontispiece. Crown 8vo. Cloth, price 3*s.* 6*d.*

GARDNER (H.).

Sunflowers. A Book of Verses. Fcap. 8vo. Cloth, price 5*s.*

GARDNER (J.), M.D.

Longevity: The Means of Prolonging Life after Middle Age. Third Edition, revised and enlarged. Small crown 8vo. Cloth, price 4*s.*

GARRETT (E.).

By Still Waters. A Story for Quiet Hours. With Seven Illustrations. Crown 8vo. Cloth, price 6*s.*

GIBBON (Charles).

For Lack of Gold. With a Frontispiece. Crown 8vo. Cloth, price 3*s.* 6*d.*

Robin Gray. With a Frontispiece. Crown 8vo. Cloth, price 3*s.* 6*d.*

GILBERT (Mrs.).

Autobiography and other Memorials. Edited by Josiah Gilbert. Second Edition. In 2 vols. With 2 Steel Portraits and several Wood Engravings. Post 8vo. Cloth, price 24*s.*

GILL (Rev. W. W.), B.A.

Myths and Songs from the South Pacific. With a Preface by F. Max Müller, M.A., Professor of Comparative Philology at Oxford. Post 8vo. Cloth, price 9*s.*

GODKIN (James).

The Religious History of Ireland: Primitive, Papal, and Protestant. Including the Evangelical Missions, Catholic Agitations, and Church Progress of the last half Century. 8vo. Cloth, price 12*s.*

GODWIN (William).

William Godwin: His Friends and Contemporaries. With Portraits and Facsimiles of the handwriting of Godwin and his Wife. By C. Kegan Paul. 2 vols. Demy 8vo. Cloth, price 28*s.*

The Genius of Christianity Unveiled. Being. Essays never before published. Edited, with a Preface, by C. Kegan Paul. Crown 8vo. Cloth, price 7*s.* 6*d.*

GOETZE (Capt. A. von).

Operations of the German Engineers during the War of 1870-1871. Published by Authority, and in accordance with Official Documents. Translated from the German by Colonel G. Graham, V.C., C.B., R.E. With 6 large Maps. Demy 8vo. Cloth, price 21*s.*

GOODENOUGH (Commodore J. G.), R.N., C.B., C.M.G.

Journals of, during his Last Command as Senior Officer on the Australian Station, 1873-1875. Edited, with a Memoir, by his Widow. With Maps, Woodcuts, and Steel Engraved Portrait. Square post 8vo. Cloth, price 14*s.*

GOODMAN (W.).

Cuba, the Pearl of the Antilles. Crown 8vo. Cloth, price 7*s.* 6*d.*

GOULD (Rev. S. Baring), M.A.

The Vicar of Morwenstow: a Memoir of the Rev. R. S. Hawker. With Portrait. Third Edition, revised. Square post 8vo. Cloth, 10*s.* 6*d.*

GRANVILLE (A. B.), M.D., F.R.S., &c.

Autobiography of A. B. Granville, F.R.S., etc. Edited, with a brief account of the concluding years of his life, by his youngest Daughter, Paulina B. Granville. 2 vols. With a Portrait. Second Edition. Demy 8vo. Cloth, price 32*s.*

GRAY (Mrs. Russell).

Lisette's Venture. A Novel. 2 vols. Crown 8vo. Cloth.

GREEN (T. Bowden).

Fragments of Thought. Dedicated by permission to the Poet Laureate. Crown 8vo. Cloth, price 7*s.* 6*d.*

GREENWOOD (J.), "The Amateur Casual."

In Strange Company; or, The Note Book of a Roving Correspondent. Second Edition. Crown 8vo. Cloth, price 6*s.*

GREY (John), of Dilston.

John Grey (of Dilston): Memoirs. By Josephine E. Butler. New and Revised Edition. Crown 8vo. Cloth, price 3*s.* 6*d.*

GRIFFITH (Rev. T.), A.M.

Studies of the Divine Master. Demy 8vo. Cloth, price 12*s.*

GRIFFITHS (Capt. Arthur).

Memorials of Millbank, and Chapters in Prison History. With Illustrations by R. Goff and the Author. 2 vols. Post 8vo. Cloth, price 21*s.*

The Queen's Shilling. A Novel. 2 vols. Cloth.

GRIMLEY (Rev. H. N.), M.A., Professor of Mathematics in the University College of Wales, and Chaplain of Tremadoc Church.

Tremadoc Sermons, chiefly on the SPIRITUAL BODY, the UNSEEN WORLD, and the DIVINE HUMANITY. Crown 8vo. Cloth, price 7s. 6d.

GRÜNER (M. L.).

Studies of Blast Furnace Phenomena. Translated by L. D. B. Gordon, F.R.S.E., F.G.S. Demy 8vo. Cloth, price 7s. 6d.

GURNEY (Rev. A. T.).

Words of Faith and Cheer. A Mission of Instruction and Suggestion. Crown 8vo. Cloth, price 6s.

First Principles in Church and State. Demy 8vo. Sewed, price 1s. 6d.

HAECKEL (Prof. Ernst).

The History of Creation. Translation revised by Professor E. Ray Lankester, M.A., F.R.S. With Coloured Plates and Genealogical Trees of the various groups of both plants and animals. 2 vols. Second Edition. Post 8vo. Cloth, price 32s.

HARCOURT (Capt. A. F. P.).

The Shakespeare Argosy. Containing much of the wealth of Shakespeare's Wisdom and Wit, alphabetically arranged and classified. Crown 8vo. Cloth, price 6s.

HAWEIS (Rev. H. R.), M.A.

Speech in Season. Third Edition. Crown 8vo. Cloth, price 9s.

Thoughts for the Times. Ninth Edition. Crown 8vo. Cloth, price 7s. 6d.

Unsectarian Family Prayers, for Morning and Evening for a Week, with short selected passages from the Bible. Square crown 8vo. Cloth, price 3s. 6d.

HAWTHORNE (Julian).

Bressant. A Romance. 2 vols. Crown 8vo. Cloth.

Idolatry. A Romance. 2 vols. Crown 8vo. Cloth.

HAWTHORNE (Nathaniel).

Nathaniel Hawthorne. A Memoir with Stories, now first published in this country. By H. A. Page. Post 8vo. Cloth, price 7s. 6d.

Septimius. A Romance. Second Edition. Crown 8vo. Cloth, price 9s.

HAYMAN (H.), D.D., late Head Master of Rugby School.

Rugby School Sermons. With an Introductory Essay on the Indwelling of the Holy Spirit. Crown 8vo. Cloth, price 7s. 6d.

Heathergate. A Story of Scottish Life and Character. By a New Author. 2 vols. Crown 8vo. Cloth.

HELLWALD (Baron F. von).

The Russians in Central Asia. A Critical Examination, down to the present time, of the Geography and History of Central Asia. Translated by Lieut.-Col. Theodore Wirgman, LL.B. Large post 8vo. With Map. Cloth, price 12s.

HELVIG (Capt. H.).

The Operations of the Bavarian Army Corps. Translated by Captain G. S. Schwabe. With Five large Maps. In 2 vols. Demy 8vo. Cloth, price 24s.

HINTON (James).

The Place of the Physician. To which is added ESSAYS ON THE LAW OF HUMAN LIFE, AND ON THE RELATION BETWEEN ORGANIC AND INORGANIC WORLDS. Second Edition. Crown 8vo. Cloth, price 3s. 6d.

Physiology for Practical Use. By various Writers. With 50 Illustrations. 2 vols. Second Edition. Crown 8vo. Cloth, price 12s. 6d.

HINTON (James)—*continued:*

An Atlas of Diseases of the Membrana Tympani. With Descriptive Text. Post 8vo. Price £6 6*s.*

The Questions of Aural Surgery. With Illustrations. 2 vols. Post 8vo. Cloth, price 12*s.* 6*d.*

H. J. C.

The Art of Furnishing. A Popular Treatise on the Principles of Furnishing, based on the Laws of Common Sense, Requirement, and Picturesque Effect. Small crown 8vo. Cloth, price 3*s.* 6*d.*

HOCKLEY (W. B.).

Tales of the Zenana; or, A Nuwab's Leisure Hours. By the Author of "Pandurang Hari." With a Preface by Lord Stanley of Alderley. 2 vols. Crown 8vo. Cloth, price 21*s.*

Pandurang Hari; or, Memoirs of a Hindoo. A Tale of Mahratta Life sixty years ago. With a Preface by Sir H. Bartle E. Frere, G.C.S.I., &c. 2 vols. Crown 8vo. Cloth, price 21*s.*

HOFFBAUER (Capt.).

The German Artillery in the Battles near Metz. Based on the official reports of the German Artillery. Translated by Capt. E. O. Hollist. With Map and Plans. Demy 8vo. Cloth, price 21*s.*

Hogan, M.P.

A Novel. 3 vols. Crown 8vo. Cloth.

HOLMES (E. G. A.).

Poems. Fcap. 8vo. Cloth, price 5*s.*

HOLROYD (Major W. R. M.)

Tas-hil ul Kālām; or, Hindustani made Easy. Crown 8vo. Cloth, price 5*s.*

HOPE (James L. A.).

In Quest of Coolies. With Illustrations. Second Edition. Crown 8vo. Cloth, price 6*s.*

HOOPER (Mary).

Little Dinners: How to Serve them with Elegance and Economy. Eleventh Edition. Crown 8vo. Cloth, price 5*s.*

Cookery for Invalids, Persons of Delicate Digestion, and Children. Crown 8vo. Cloth, price 3*s.* 6*d.*

HOOPER (Mrs. G.).

The House of Raby. With a Frontispiece. Crown 8vo. Cloth, price 3*s.* 6*d.*

HOPKINS (M.).

The Port of Refuge; or, Counsel and Aid to Shipmasters in Difficulty, Doubt, or Distress. Crown 8vo. Cloth, price 6*s.*

HORNE (William), M.A.

Reason and Revelation: an Examination into the Nature and Contents of Scripture Revelation, as compared with other Forms of Truth. Demy 8vo. Cloth, price 12*s.*

HOWARD (Mary M.).

Beatrice Aylmer, and other Tales. Crown 8vo. Cloth, price 6*s.*

HOWARD (Rev. G. B.).

An Old Legend of St. Paul's. Fcap. 8vo. Cloth, price 4*s.* 6*d.*

HOWELL (James).

A Tale of the Sea, Sonnets, and other Poems. Fcap. 8vo. Cloth, price 5*s.*

HUGHES (Allison).

Penelope and other Poems. Fcap. 8vo. Cloth, price 4*s.* 6*d.*

HULL (Edmund C. P.).

The European in India. With a Medical Guide for Anglo-Indians. By R. R. S. Mair, M.D., F.R.C.S.E. Second Edition, Revised and Corrected. Post 8vo. Cloth, price 6*s.*

HUMPHREY (Rev. W.).

Mr. Fitzjames Stephen and Cardinal Bellarmine. Demy 8vo. Sewed, price 1*s.*

HUTTON (James).

Missionary Life in the Southern Seas. With Illustrations. Crown 8vo. Cloth, price 7*s.* 6*d.*

IGNOTUS.

Culmshire Folk. A Novel. New and Cheaper Edition. Crown 8vo. Cloth, price 6*s.*

INGELOW (Jean).

The Little Wonder-horn. A Second Series of "Stories Told to a Child." With Fifteen Illustrations. Square 24mo. Cloth, price 3*s.* 6*d.*

Off the Skelligs. (Her First Romance.) 4 vols. Crown 8vo. Cloth.

International Scientific Series (The).

I. The Forms of Water in Clouds and Rivers, Ice and Glaciers. By J. Tyndall, LL.D., F.R.S. With 25 Illustrations. Sixth Edition. Crown 8vo. Cloth, price 5*s.*

II. Physics and Politics; or, Thoughts on the Application of the Principles of "Natural Selection" and "Inheritance" to Political Society. By Walter Bagehot. Third Edition. Crown 8vo. Cloth, price 4*s.*

III. Foods. By Edward Smith, M.D., LL.B., F.R.S. With numerous Illustrations. Fourth Edition. Crown 8vo. Cloth, price 5*s.*

IV. Mind and Body: The Theories of their Relation. By Alexander Bain, LL.D. With Four Illustrations. Fifth Edition. Crown 8vo. Cloth, price 4*s.*

V. The Study of Sociology. By Herbert Spencer. Fifth Edition. Crown 8vo. Cloth, price 5*s.*

VI. On the Conservation of Energy. By Balfour Stewart, M.A., LL.D., F.R.S. With 14 Illustrations. Third Edition. Crown 8vo. Cloth, price 5*s.*

International Scientific Series (The)—*continued.*

VII. Animal Locomotion; or, Walking, Swimming, and Flying. By J. B. Pettigrew, M.D., F.R.S., etc. With 130 Illustrations. Second Edition. Crown 8vo. Cloth, price 5*s.*

VIII. Responsibility in Mental Disease. By Henry Maudsley, M.D. Second Edition. Crown 8vo. Cloth, price 5*s.*

IX. The New Chemistry. By Professor J. P. Cooke, of the Harvard University. With 31 Illustrations. Third Edition. Crown 8vo. Cloth, price 5*s.*

X. The Science of Law. By Professor Sheldon Amos. Second Edition. Crown 8vo. Cloth, price 5*s.*

XI. Animal Mechanism. A Treatise on Terrestrial and Aerial Locomotion. By Professor E. J. Marey. With 117 Illustrations. Second Edition. Crown 8vo. Cloth, price 5*s.*

XII. The Doctrine of Descent and Darwinism. By Professor Oscar Schmidt (Strasburg University). With 26 Illustrations. Third Edition. Crown 8vo. Cloth, price 5*s.*

XIII. The History of the Conflict between Religion and Science. By J. W. Draper, M.D., LL.D. Seventh Edition. Crown 8vo. Cloth, price 5*s.*

XIV. Fungi; their Nature, Influences, Uses, &c. By M. C. Cooke, M.A., LL.D. Edited by the Rev. M. J. Berkeley, M.A., F.L.S. With numerous Illustrations. Second Edition. Crown 8vo. Cloth, price 5*s.*

XV. The Chemical Effects of Light and Photography. By Dr. Hermann Vogel (Polytechnic Academy of Berlin). Translation thoroughly revised. With 100 Illustrations. Third Edition. Crown 8vo. Cloth, price 5*s.*

XVI. The Life and Growth of Language. By William Dwight Whitney, Professor of Sanskrit and Comparative Philology in Yale College, New Haven. Second Edition. Crown 8vo. Cloth, price 5*s.*

International Scientific Series (The)—*continued.*

XVII. Money and the Mechanism of Exchange. By W. Stanley Jevons, M.A., F.R.S. Third Edition. Crown 8vo. Cloth, price 5*s*.

XVIII. The Nature of Light: With a General Account of Physical Optics. By Dr. Eugene Lommel, Professor of Physics in the University of Erlangen. With 188 Illustrations and a table of Spectra in Chromolithography. Second Edition. Crown 8vo. Cloth, price 5*s*.

XIX. Animal Parasites and Messmates. By Monsieur Van Beneden, Professor of the University of Louvain, Correspondent of the Institute of France. With 83 Illustrations. Second Edition. Crown 8vo. Cloth, price 5*s*.

XX. Fermentation. By Professor Schützenberger, Director of the Chemical Laboratory at the Sorbonne. With 28 Illustrations. Second Edition. Crown 8vo. Cloth, price 5*s*.

XXI. The Five Senses of Man. By Professor Bernstein, of the University of Halle. With 91 Illustrations. Second Edition. Crown 8vo. Cloth, price 5*s*.

XXII. The Theory of Sound in its Relation to Music. By Professor Pietro Blaserna, of the Royal University of Rome. With numerous Illustrations. Crown 8vo. Cloth, price 5*s*.

Forthcoming Volumes.

Prof. W. Kingdon Clifford, M.A. The First Principles of the Exact Sciences explained to the Non-mathematical.

Prof. T. H. Huxley, LL.D., F.R.S. Bodily Motion and Consciousness.

Dr. W. B. Carpenter, LL.D., F.R.S. The Physical Geography of the Sea.

W. Lauder Lindsay, M.D., F.R.S.E. Mind in the Lower Animals.

Sir John Lubbock, Bart., F.R.S. On Ants and Bees.

Prof. W. T. Thiselton Dyer, B.A., B.Sc. Form and Habit in Flowering Plants.

International Scientific Series (The)—*continued.*

Mr. J. N. Lockyer, F.R.S. Spectrum Analysis.

Prof. Michael Foster, M.D. Protoplasm and the Cell Theory.

H. Charlton Bastian, M.D., F.R.S. The Brain as an Organ of Mind.

Prof. A. C. Ramsay, LL.D., F.R.S. Earth Sculpture: Hills, Valleys, Mountains, Plains, Rivers, Lakes; how they were Produced, and how they have been Destroyed.

Prof. J. Rosenthal. General Physiology of Muscles and Nerves.

P. Bert (Professor of Physiology, Paris). Forms of Life and other Cosmical Conditions.

Prof. Corfield, M.A., M.D. (Oxon.) Air in its relation to Health.

JACKSON (T. G.).

Modern Gothic Architecture. Crown 8vo. Cloth, price 5*s*.

JACOB (Maj.-Gen. Sir G. Le Grand), K.C.S.I., C.B.

Western India Before and during the Mutinies. Pictures drawn from life. Second Edition. Crown 8vo. Cloth, price 7*s*. 6*d*.

JENKINS (E.) and RAYMOND (J.), Esqs.

A Legal Handbook for Architects, Builders, and Building Owners. Second Edition Revised. Crown 8vo. Cloth, price 6*s*.

JENKINS (Rev. R. C.), M.A.

The Privilege of Peter and the Claims of the Roman Church confronted with the Scriptures, the Councils, and the Testimony of the Popes themselves. Fcap. 8vo. Cloth, price 3*s*. 6*d*.

JENNINGS (Mrs. Vaughan).

Rahel: Her Life and Letters. With a Portrait from the Painting by Daffinger. Square post 8vo. Cloth, price 7*s*. 6*d*.

JEVONS (W. Stanley), M.A., F.R.S.

Money and the Mechanism of Exchange. Second Edition. Crown 8vo. Cloth, price 5*s*.
Volume XVII. of The International Scientific Series.

KAUFMANN (Rev. M.), B.A.

Socialism: Its Nature, its Dangers, and its Remedies considered. Crown 8vo. Cloth, price 7*s*. 6*d*.

KEATINGE (Mrs.).

Honor Blake: The Story of a Plain Woman. 2 vols. Crown 8vo. Cloth.

KER (David).

On the Road to Khiva. Illustrated with Photographs of the Country and its Inhabitants, and a copy of the Official Map in use during the Campaign, from the Survey of Captain Leusilin. Post 8vo. Cloth, price 12*s*.

The Boy Slave in Bokhara. A Tale of Central Asia. With Illustrations. Crown 8vo. Cloth, price 5*s*.

The Wild Horseman of the Pampas. Illustrated. Crown 8vo. Cloth, price 5*s*.

KING (Alice).

A Cluster of Lives. Crown 8vo. Cloth, price 7*s*. 6*d*.

KING (Mrs. Hamilton).

The Disciples. A New Poem. Second Edition, with some Notes. Crown 8vo. Cloth, price 7*s*. 6*d*.

Aspromonte, and other Poems. Second Edition. Fcap. 8vo. Cloth, price 4*s*. 6*d*.

KINGSFORD (Rev. F. W.), M.A., Vicar of St. Thomas's, Stamford Hill; late Chaplain H. E. I. C. (Bengal Presidency).

Hartham Conferences; or, Discussions upon some of the Religious Topics of the Day. "Audi alteram partem." Crown 8vo. Cloth, price 3*s*. 6*d*.

KNIGHT (A. F. C.).

Poems. Fcap 8vo. Cloth, price 5*s*.

LACORDAIRE (Rev. Père).

Life: Conferences delivered at Toulouse. A New and Cheaper Edition. Crown 8vo. Cloth, price 3*s*. 6*d*.

Lady of Lipari (The).

A Poem in Three Cantos. Fcap. 8vo. Cloth, price 5*s*.

LAURIE (J. S.).

Educational Course of Secular School Books for India:

The First Hindustani Reader. Stiff linen wrapper, price 6*d*.

The Second Hindustani Reader. Stiff linen wrapper, price 6*d*.

The Oriental (English) Reader. Book I., price 6*d*.; II., price 7½*d*.; III., price 9*d*.; IV., price 1*s*.

Geography of India; with Maps and Historical Appendix, tracing the Growth of the British Empire in Hindustan. Fcap. 8vo. Cloth, price 1*s*. 6*d*.

LAYMANN (Capt.).

The Frontal Attack of Infantry. Translated by Colonel Edward Newdigate. Crown 8vo. Cloth, price 2*s*. 6*d*.

L. D. S.

Letters from China and Japan. With Illustrated Title-page. Crown 8vo. Cloth, price 7*s*. 6*d*.

LEANDER (Richard).

Fantastic Stories. Translated from the German by Paulina B. Granville. With Eight full-page Illustrations by M. E. Fraser-Tytler. Crown 8vo. Cloth, price 5*s*.

LEATHES (Rev. S.), M.A.

The Gospel Its Own Witness. Crown 8vo. Cloth, price 5*s*.

LEE (Rev. F. G.), D.C.L.

The Other World; or, Glimpses of the Supernatural. 2 vols. A New Edition. Crown 8vo. Cloth, price 15*s*.

LEE (Holme).

Her Title of Honour. A Book for Girls. New Edition. With a Frontispiece. Crown 8vo. Cloth, price 5*s*.

LENOIR (J.).

Fayoum; or, Artists in Egypt. A Tour with M. Gérome and others. With 13 Illustrations. A New and Cheaper Edition. Crown 8vo. Cloth, price 3*s*. 6*d*.

Leonora Christina, Memoirs of, Daughter of Christian IV. of Denmark. Written during her Imprisonment in the Blue Tower of the Royal Palace at Copenhagen, 1663-1685. Translated by F. E. BUNNETT. With an Autotype Portrait of the Princess. A New and Cheaper Edition. Medium 8vo. Cloth, price 5*s*.

LEWIS (Mary A.).

A Rat with Three Tales. With Four Illustrations by Catherine F. Frere. Cloth, price 5*s*.

LISTADO (J. T.).

Civil Service. A Novel. 2 vols. Crown 8vo. Cloth.

LOCKER (F.).

London Lyrics. A New and Revised Edition, with Additions and a Portrait of the Author. Crown 8vo. Cloth, elegant, price 7*s*. 6*d*.

LOMMEL (Dr. E.).

The Nature of Light: With a General Account of Physical Optics. Second Edition. With 188 Illustrations and a Table of Spectra in Chromolithography. Crown 8vo. Cloth, price 5*s*.

Volume XVIII. of The International Scientific Series.

LORIMER (Peter), D.D.

John Knox and the Church of England: His Work in her Pulpit, and his Influence upon her Liturgy, Articles, and Parties. Demy 8vo. Cloth, price 12*s*.

LOTHIAN (Roxburghe).

Dante and Beatrice from 1282 to 1290. A Romance. 2 vols. Post 8vo. Cloth, price 24*s*.

LOVEL (Edward).

The Owl's Nest in the City: A Story. Crown 8vo. Cloth.

LOVER (Samuel), R.H.A.

The Life of Samuel Lover, R. H. A.; Artistic, Literary, and Musical. With Selections from his Unpublished Papers and Correspondence. By Bayle Bernard. 2 vols. With a Portrait. Post 8vo. Cloth, price 21*s*.

LOWER (M. A.), M.A., F.S.A.

Wayside Notes in Scandinavia. Being Notes of Travel in the North of Europe. Crown 8vo. Cloth, price 9*s*.

LUCAS (Alice).

Translations from the Works of German Poets of the 18th and 19th Centuries. Fcap. 8vo. Cloth, price 5*s*.

LYONS (R. T.), Surg.-Maj. Bengal Army.

A Treatise on Relapsing Fever. Post 8vo. Cloth, price 7*s*. 6*d*.

MACAULAY (J.), M.A., M.D., Edin.

The Truth about Ireland: Tours of Observation in 1872 and 1875. With Remarks on Irish Public Questions. Being a Second Edition of "Ireland in 1872," with a New and Supplementary Preface. Crown 8vo. Cloth, price 3*s*. 6*d*.

MAC DONALD (G.).

Malcolm. A Novel. 3 vols. Second Edition. Crown 8vo. Cloth.

St. George and St. Michael. 3 vols. Crown 8vo. Cloth.

MACLACHLAN (A. N. C.), M.A.

William Augustus, Duke of Cumberland: being a Sketch of his Military Life and Character, chiefly as exhibited in the General Orders of His Royal Highness, 1745—1747. With Illustrations. Post 8vo. Cloth, price 15*s*.

MAC KENNA (S. J.).

Plucky Fellows. A Book for Boys. With Six Illustrations. Second Edition. Crown 8vo. Cloth, price 3*s*. 6*d*.

At School with an Old Dragoon. With Six Illustrations. Second Edition. Crown 8vo. Cloth, price 5*s*.

MAIR (R. S.), M.D., F.R.C.S.E.

The Medical Guide for Anglo-Indians. Being a Compendium of Advice to Europeans in India, relating to the Preservation and Regulation of Health. With a Supplement on the Management of Children in India. Crown 8vo. Limp cloth, price 3*s*. 6*d*.

MANNING (His Eminence Cardinal).

Essays on Religion and Literature. By various Writers. Third Series. Demy 8vo. Cloth, price 10*s*. 6*d*.

MAREY (E. J.).

Animal Mechanics. A Treatise on Terrestrial and Aerial Locomotion. With 117 Illustrations. Second Edition. Crown 8vo. Cloth, price 5*s*.

Volume XI. of The International Scientific Series.

MARKEWITCH (B.).

The Neglected Question. Translated from the Russian, by the Princess Ourousoff, and dedicated by Express Permission to Her Imperial and Royal Highness Marie Alexandrovna, the Duchess of Edinburgh. 2 vols. Crown 8vo. Cloth, price 14*s*.

MARRIOTT (Maj.-Gen. W. F.), C.S.I.

A Grammar of Political Economy. Crown 8vo. Cloth, price 6*s*.

MARSHALL (H.).

The Story of Sir Edward's Wife. A Novel. Crown 8vo. Cloth, price 10*s*. 6*d*.

MASTERMAN (J.).

Half-a-dozen Daughters. With a Frontispiece. Crown 8vo. Cloth, price 3*s*. 6*d*.

MAUDSLEY (Dr. H.).

Responsibility in Mental Disease. Second Edition. Crown 8vo. Cloth, price 5*s*.

Volume VIII. of The International Scientific Series.

MAUGHAN (W. C.).

The Alps of Arabia; or, Travels through Egypt, Sinai, Arabia, and the Holy Land. With Map. Second Edition. Demy 8vo. Cloth, price 5*s*.

MAURICE (C. E.).

Lives of English Popular Leaders. No. 1.—STEPHEN LANGTON. Crown 8vo. Cloth, price 7*s*. 6*d*. No. 2.—TYLER, BALL, and OLDCASTLE. Crown 8vo. Cloth, price 7*s*. 6*d*.

Mazzini (Joseph).

A Memoir. By E. A. V. Two Photographic Portraits. Crown 8vo. Cloth, price 3*s*. 6*d*.

MEDLEY (Lieut.-Col. J. G.), R.E.

An Autumn Tour in the United States and Canada. Crown 8vo. Cloth, price 5*s*.

MENZIES (Sutherland).

Memoirs of Distinguished Women. 2 vols. Post 8vo. Cloth, price 10*s*. 6*d*.

MICKLETHWAITE (J. T.), F.S.A.

Modern Parish Churches: Their Plan, Design, and Furniture. Crown 8vo. Cloth, price 7*s*. 6*d*.

MILNE (James).

Tables of Exchange for the Conversion of Sterling Money into Indian and Ceylon Currency, at Rates from 1*s*. 8*d*. to 2*s*. 3*d*. per Rupee. Second Edition. Demy 8vo. Cloth, price £2 2*s*.

MIRUS (Maj.-Gen. von).

Cavalry Field Duty. Translated by Major Frank S. Russell, 14th (King's) Hussars. Crown 8vo. Cloth limp, price 7*s.* 6*d.*

MIVART (St. George), F.R.S.

Contemporary Evolution: An Essay on some recent Social Changes. Post 8vo. Cloth, price 7*s.* 6*d.*

MOORE (Rev. D.), M.A.

Christ and His Church. By the Author of "The Age and the Gospel," &c. Crown 8vo. Cloth, price 3*s.* 6*d.*

MOORE (Rev. T.).

Sermonettes: on Synonymous Texts, taken from the Bible and Book of Common Prayer, for the Study, Family Reading, and Private Devotion. Small crown 8vo. Cloth, price 4*s.* 6*d.*

MORELL (J. R.).

Euclid Simplified in Method and Language. Being a Manual of Geometry. Compiled from the most important French Works, approved by the University of Paris and the Minister of Public Instruction. Fcap. 8vo. Cloth, price 2*s.* 6*d.*

MORICE (Rev. F. D.), M.A.

The Olympian and Pythian Odes of Pindar. A New Translation in English Verse. Crown 8vo. Cloth, price 7*s.* 6*d.*

MORLEY (Susan).

Aileen Ferrers. A Novel. 2 vols. Crown 8vo. Cloth.

Throstlethwaite. A Novel. 3 vols. Crown 8vo. Cloth.

MORSE (E. S.), Ph.D.

First Book of Zoology. With numerous Illustrations. Crown 8vo. Cloth, price 5*s.*

MOSTYN (Sydney).

Perplexity. A Novel. 3 vols. Crown 8vo. Cloth.

MUSGRAVE (Anthony).

Studies in Political Economy. Crown 8vo. Cloth, price 6*s.*

My Sister Rosalind.

A Novel. By the Author of "Christiana North," and "Under the Limes." 2 vols. Cloth.

NAAKÉ (J. T.).

Slavonic Fairy Tales. From Russian, Servian, Polish, and Bohemian Sources. With Four Illustrations. Crown 8vo. Cloth, price 5*s.*

NEWMAN (J. H.), D.D.

Characteristics from the Writings of. Being Selections from his various Works. Arranged with the Author's personal approval. Second Edition. With Portrait. Crown 8vo. Cloth, price 6*s.*

*** A Portrait of the late Rev. Dr. J. H. Newman, mounted for framing, can be had, price 2*s.* 6*d.*

NEWMAN (Mrs.).

Too Late. A Novel. 2 vols. Crown 8vo. Cloth.

NEW WRITER (A).

Songs of Two Worlds. By a New Writer. Third Series. Second Edition. Fcap. 8vo. Cloth, price 5*s.*

The Epic of Hades. Fcap. 8vo. Cloth, price 5*s.*

NOBLE (J. A.).

The Pelican Papers. Reminiscences and Remains of a Dweller in the Wilderness. Crown 8vo. Cloth, price 6*s.*

NORMAN PEOPLE (The).

The Norman People, and their Existing Descendants in the British Dominions and the United States of America. Demy 8vo. Cloth, price 21*s.*

NORRIS (Rev. Alfred).

The Inner and Outer Life Poems. Fcap. 8vo. Cloth, price 6*s*.

NOTREGE (John), A.M.

The Spiritual Function of a Presbyter in the Church of England. Crown 8vo. Cloth, red edges, price 3*s*. 6*d*.

Oriental Sporting Magazine (The).

A Reprint of the first 5 Volumes, in 2 Volumes. Demy 8vo. Cloth, price 28*s*.

Our Increasing Military Difficulty, and one Way of Meeting it. Demy 8vo. Stitched, price 1*s*.

PAGE (Capt. S. F.).

Discipline and Drill. Cheaper Edition. Crown 8vo. Price 1*s*.

PALGRAVE (W. Gifford).

Hermann Agha. An Eastern Narrative. 2 vols. Crown 8vo. Cloth, extra gilt, price 18*s*.

PANDURANG HARI;

Or Memoirs of a Hindoo. With an Introductory Preface by Sir H. Bartle E. Frere, G.C.S.I., C.B. 2 vols. Crown 8vo. Cloth, price 21*s*.

PARKER (Joseph), D.D.

The Paraclete: An Essay on the Personality and Ministry of the Holy Ghost, with some reference to current discussions. Second Edition. Demy 8vo. Cloth, price 12*s*.

PARR (Harriet).

Echoes of a Famous Year. Crown 8vo. Cloth, price 8*s*. 6*d*.

PAUL (C. Kegan).

Goethe's Faust. A New Translation in Rime. Crown 8vo. Cloth, price 6*s*.

William Godwin: His Friends and Contemporaries. With Portraits and Facsimiles of the Handwriting of Godwin and his Wife. 2 vols. Square post 8vo. Cloth, price 28*s*.

PAUL (C. Kegan).

The Genius of Christianity Unveiled. Being Essays never before published. Edited, with a Preface, by C. Kegan Paul. Crown 8vo. Cloth, price 7*s*. 6*d*.

PAYNE (John).

Songs of Life and Death. Crown 8vo. Cloth, price 5*s*.

PAYNE (Prof.).

Lectures on Education. Price 6*d*. each.

I. Pestalozzi: the Influence of His Principles and Practice.
II. Fröbel and the Kindergarten System. Second Edition.
III. The Science and Art of Education.
IV. The True Foundation of Science Teaching.

A Visit to German Schools: Elementary Schools in Germany. Notes of a Professional Tour to inspect some of the Kindergartens, Primary Schools, Public Girls' Schools, and Schools for Technical Instruction in Hamburgh, Berlin, Dresden, Weimar, Gotha, Eisenach, in the autumn of 1874. With Critical Discussions of the General Principles and Practice of Kindergartens and other Schemes of Elementary Education. Crown 8vo. Cloth, price 4*s*. 6*d*.

PELLETAN (E.).

The Desert Pastor, Jean Jarousseau. Translated from the French. By Colonel E. P. De L'Hoste. With a Frontispiece. New Edition. Fcap. 8vo. Cloth, price 3*s*. 6*d*.

PENRICE (Maj. J.), B.A.

A Dictionary and Glossary of the Ko-ran. With copious Grammatical References and Explanations of the Text. 4to. Cloth, price 21*s*.

PERCEVAL (Rev. P.).

Tamil Proverbs, with their English Translation. Containing upwards of Six Thousand Proverbs. Third Edition. Demy 8vo. Sewed, price 9*s*.

PERRIER (A.).

A Winter in Morocco. With Four Illustrations. A New and Cheaper Edition. Crown 8vo. Cloth, price 3*s*. 6*d*.

A Good Match. A Novel. 2 vols. Crown 8vo. Cloth.

PERRY (Rev. S. J.), F.R.S.

Notes of a Voyage to Kerguelen Island, to observe the Transit of Venus. Demy 8vo. Sewed, price 2*s*.

PESCHEL (Dr. Oscar).

The Races of Man and their Geographical Distribution. Large crown 8vo. Cloth, price 9*s*.

PETTIGREW (J. Bell), M.D., F.R.S.

Animal Locomotion; or, Walking, Swimming, and Flying. With 130 Illustrations. Second Edition. Crown 8vo. Cloth, price 5*s*.
Volume VII. of The International Scientific Series.

PIGGOT (J.), F.S.A., F.R.G.S.

Persia—Ancient and Modern. Post 8vo. Cloth, price 10*s*. 6*d*.

POUSHKIN (A. S.).

Russian Romance. Translated from the Tales of Belkin, etc. By Mrs. J. Buchan Telfer (*née* Mouravieff). Crown 8vo. Cloth, price 7*s*. 6*d*.

POWER (H.).

Our Invalids: How shall we Employ and Amuse Them? Fcap. 8vo. Cloth, price 2*s*. 6*d*.

POWLETT (Lieut. N.), R.A.

Eastern Legends and Stories in English Verse. Crown 8vo. Cloth, price 5*s*.

PRESBYTER.

Unfoldings of Christian Hope. An Essay showing that the Doctrine contained in the Damnatory Clauses of the Creed commonly called Athanasian is unscriptural. Small crown 8vo. Cloth, price 4*s*. 6*d*.

PRICE (Prof. Bonamy).

Currency and Banking. Crown 8vo. Cloth, price 6*s*.

PROCTOR (Richard A.), B.A.

Our Place among Infinities. A Series of Essays contrasting our little abode in space and time with the Infinities around us. To which are added Essays on "Astrology," and "The Jewish Sabbath." Second Edition. Crown 8vo. Cloth, price 6*s*.

The Expanse of Heaven. A Series of Essays on the Wonders of the Firmament. With a Frontispiece. Second Edition. Crown 8vo. Cloth, price 6*s*.

PUBLIC SCHOOLBOY.

The Volunteer, the Militiaman, and the Regular Soldier. Crown 8vo. Cloth, price 5*s*.

RANKING (B. M.).

Streams from Hidden Sources. Crown 8vo. Cloth, price 6*s*.

Ready-Money Mortiboy. A Matter-of-Fact Story. With Frontispiece. Crown 8vo. Cloth, price 3*s*. 6*d*.

REANEY (Mrs. G. S.).

Waking and Working; or, from Girlhood to Womanhood. With a Frontispiece. Crown 8vo. Cloth, price 5*s*.

Sunbeam Willie, and other Stories. Three Illustrations. Royal 16mo. Cloth, price 1*s*. 6*d*.

Reginald Bramble. A Cynic of the Nineteenth Century. An Autobiography. Crown 8vo. Cloth, price 10*s*. 6*d*.

REID (T. Wemyss).

Cabinet Portraits. Biographical Sketches of Statesmen of the Day. Crown 8vo. Cloth, price 7*s*. 6*d*.

RHOADES (James).

Timoleon. A Dramatic Poem. Fcap. 8vo. Cloth, price 5*s*.

RIBOT (Prof. Th.).

Contemporary English Psychology. Second Edition. A Revised and Corrected Translation from the latest French Edition. Large post 8vo. Cloth, price 9*s.*

Heredity: A Psychological Study on its Phenomena, its Laws, its Causes, and its Consequences. Large crown 8vo. Cloth, price 9*s.*

ROBERTSON (The Late Rev. F. W.), M.A., of Brighton.

The Late Rev. F. W. Robertson, M.A., Life and Letters of. Edited by the Rev. Stopford Brooke, M.A., Chaplain in Ordinary to the Queen.
I. 2 vols., uniform with the Sermons. With Steel Portrait. Crown 8vo. Cloth, price 7*s.* 6*d.*
II. Library Edition, in Demy 8vo., with Two Steel Portraits. Cloth, price 12*s.*
III. A Popular Edition, in 1 vol. Crown 8vo. Cloth, price 6*s.*

New and Cheaper Editions:—

Sermons. Four Series. Small crown 8vo. Cloth, price 3*s.* 6*d.* each.

Expository Lectures on St. Paul's Epistles to the Corinthians. A New Edition. Small crown 8vo. Cloth, price 5*s.*

Lectures and Addresses, with other literary remains. A New Edition. Crown 8vo. Cloth, price 5*s.*

An Analysis of Mr. Tennyson's "In Memoriam." (Dedicated by Permission to the Poet-Laureate.) Fcap. 8vo. Cloth, price 2*s.*

The Education of the Human Race. Translated from the German of Gotthold Ephraim Lessing. Fcap. 8vo. Cloth, price 2*s.* 6*d.*

The above Works can also be had bound in half-morocco.

*** A Portrait of the late Rev. F. W. Robertson, mounted for framing, can be had, price 2*s.* 6*d.*

ROSS (Mrs. E.), ("Nelsie Brook").

Daddy's Pet. A Sketch from Humble Life. With Six Illustrations. Royal 16mo. Cloth, price 1*s.*

RUSSELL (E. R.).

Irving as Hamlet. Second Edition. Demy 8vo. Sewed, price 1*s.*

RUSSELL (W. C.).

Memoirs of Mrs. Lætitia Boothby. Crown 8vo. Cloth, price 7*s.* 6*d.*

SADLER (S. W.), R.N.

The African Cruiser. A Midshipman's Adventures on the West Coast. With Three Illustrations. Second Edition. Crown 8vo. Cloth, price 3*s.* 6*d.*

SAMAROW (G.).

For Sceptre and Crown. A Romance of the Present Time. Translated by Fanny Wormald. 2 vols. Crown 8vo. Cloth, price 15*s.*

SAUNDERS (Katherine).

The High Mills. A Novel. 3 vols. Crown 8vo. Cloth.

Gideon's Rock, and other Stories. Crown 8vo. Cloth, price 6*s.*

Joan Merryweather, and other Stories. Crown 8vo. Cloth, price 6*s.*

Margaret and Elizabeth. A Story of the Sea. Crown 8vo. Cloth, price 6*s.*

SAUNDERS (John).

Israel Mort, Overman. A Story of the Mine. 3 vols. Crown 8vo.

Hirell. With Frontispiece. Crown 8vo. Cloth, price 3*s.* 6*d.*

Abel Drake's Wife. With Frontispiece. Crown 8vo. Cloth, price 3*s.* 6*d.*

SCHELL (Maj. von).

The Operations of the First Army under Gen. Von Goeben. Translated by Col. C. H. von Wright. Four Maps. Demy 8vo. Cloth, price 9*s.*

The Operations of the First Army under Gen. Von Steinmetz. Translated by Captain E. O. Hollist. Demy 8vo. Cloth, price 10*s.* 6*d.*

SCHERFF (Maj. W. von).

Studies in the New Infantry Tactics. Parts I. and II. Translated from the German by Colonel Lumley Graham. Demy 8vo. Cloth, price 7*s.* 6*d.*

SCHMIDT (Prof. Oscar).

The Doctrine of Descent and Darwinism. With 26 Illustrations. Third Edition. Crown 8vo. Cloth, price 5*s.*
Volume XII. of The International Scientific Series.

SCHÜTZENBERGER (Prof. F.).

Fermentation. With Numerous Illustrations. Crown 8vo. Cloth, price 5*s.*
Volume XX. of The International Scientific Series.

SCOTT (Patrick).

The Dream and the Deed, and other Poems. Fcap. 8vo. Cloth, price 5*s.*

SCOTT (W. T.).

Antiquities of an Essex Parish; or, Pages from the History of Great Dunmow. Crown 8vo. Cloth, price 5*s.* Sewed, 4*s.*

SCOTT (Robert H.).

Weather Charts and Storm Warnings. Illustrated. Crown 8vo. Cloth, price 3*s.* 6*d.*

Seeking his Fortune, and other Stories. With Four Illustrations. Crown 8vo. Cloth, price 3*s.* 6*d.*

SENIOR (N. W.).

Alexis De Tocqueville. Correspondence and Conversations with Nassau W. Senior, from 1833 to 1859. Edited by M. C. M. Simpson. 2 vols. Large post 8vo. Cloth, price 21*s.*

Journals Kept in France and Italy. From 1848 to 1852. With a Sketch of the Revolution of 1848. Edited by his Daughter, M. C. M. Simpson. 2 vols. Post 8vo. Cloth, price 24*s.*

Seven Autumn Leaves from Fairyland. Illustrated with Nine Etchings. Square crown 8vo. Cloth, price 3*s.* 6*d.*

SEYD (Ernest), F.S.S.

The Fall in the Price of Silver. Its Causes, its Consequences, and their Possible Avoidance, with Special Reference to India. Demy 8vo. Sewed, price 2*s.* 6*d.*

SHADWELL (Maj.-Gen.), C.B.

Mountain Warfare. Illustrated by the Campaign of 1799 in Switzerland. Being a Translation of the Swiss Narrative compiled from the Works of the Archduke Charles, Jomini, and others. Also of Notes by General H. Dufour on the Campaign of the Valtelline in 1635. With Appendix, Maps, and Introductory Remarks. Demy 8vo. Cloth, price 16*s.*

SHELDON (Philip).

Woman's a Riddle; or, Baby Warmstrey. A Novel. 3 vols. Crown 8vo. Cloth.

SHELLEY (Lady).

Shelley Memorials from Authentic Sources. With (now first printed) an Essay on Christianity by Percy Bysshe Shelley. With Portrait. Third Edition. Crown 8vo. Cloth, price 5*s.*

SHERMAN (Gen. W. T.).

Memoirs of General W. T. Sherman, Commander of the Federal Forces in the American Civil War. By Himself. 2 vols. With Map. Demy 8vo. Cloth, price 24*s.*
Copyright English Edition.

SHIPLEY (Rev. Orby), M.A.

Church Tracts, or Studies in Modern Problems. By various Writers. 2 vols. Crown 8vo. Cloth, price 5*s*. each.

SMEDLEY (M. B.).

Boarding-out and Pauper Schools for Girls. Crown 8vo. Cloth, price 3*s*. 6*d*.

SMITH (Edward), M.D., LL.B., F.R.S.

Health and Disease, as Influenced by the Daily, Seasonal, and other Cyclical Changes in the Human System. A New Edition. Post 8vo. Cloth, price 7*s*. 6*d*.

Foods. Profusely Illustrated. Fourth Edition. Crown 8vo. Cloth, price 5*s*.
Volume III. of The International Scientific Series.

Practical Dietary for Families, Schools, and the Labouring Classes. A New Edition. Post 8vo. Cloth, price 3*s*. 6*d*.

Tubercular Consumption in its Early and Remediable Stages. Second Edition. Crown 8vo. Cloth, price 6*s*.

SMITH (Hubert).

Tent Life with English Gipsies in Norway. With Five full-page Engravings and Thirty-one smaller Illustrations by Whymper and others, and Map of the Country showing Routes. Third Edition. Revised and Corrected. Post 8vo. Cloth, price 21*s*.

Some Time in Ireland.

A Recollection. Crown 8vo. Cloth, price 7*s*. 6*d*.

Songs for Music.

By Four Friends. Square crown 8vo. Cloth, price 5*s*.
Containing songs by Reginald A. Gatty, Stephen H. Gatty, Greville J. Chester, and Juliana H. Ewing.

SPENCER (Herbert).

The Study of Sociology. Fifth Edition. Crown 8vo. Cloth, price 5*s*.
Volume V. of The International Scientific Series.

SPICER (H.).

Otho's Death Wager. A Dark Page of History Illustrated. In Five Acts. Fcap. 8vo. Cloth, price 5*s*.

STEVENSON (Rev. W. F.).

Hymns for the Church and Home. Selected and Edited by the Rev. W. Fleming Stevenson.
The most complete Hymn Book published.
The Hymn Book consists of Three Parts:—I. For Public Worship.—II. For Family and Private Worship.—III. For Children.
**** Published in various forms and prices, the latter ranging from 8d. to 6s. Lists and full particulars will be furnished on application to the Publishers.*

STEWART (Prof. Balfour), M.A., LL.D., F.R.S.

On the Conservation of Energy. Third Edition. With Fourteen Engravings. Crown 8vo. Cloth, price 5*s*.
Volume VI. of The International Scientific Series.

STONEHEWER (Agnes).

Monacella: A Legend of North Wales. A Poem. Fcap. 8vo. Cloth, price 3*s*. 6*d*.

STRETTON (Hesba). Author of "Jessica's First Prayer."

The Storm of Life. With Ten Illustrations. Royal 16mo. Cloth, price 1*s*. 6*d*.

The Crew of the Dolphin. Illustrated. Eighth Thousand. Royal 16mo. Cloth, price 1*s*. 6*d*.

Cassy. Twenty-ninth Thousand. With Six Illustrations. Royal 16mo. Cloth, price 1*s*. 6*d*.

STRETTON (Hesba)—*continued:*

The King's Servants. Thirty-fifth Thousand. With Eight Illustrations. Royal 16mo. Cloth, price 1*s.* 6*d.*

Lost Gip. Forty-eighth Thousand. With Six Illustrations. Royal 16mo. Cloth, price 1*s.* 6*d.*
*** *Also a handsomely bound Edition, with Twelve Illustrations, price* 2*s.* 6*d.*

The Wonderful Life. Ninth Thousand. Fcap. 8vo. Cloth, price 2*s.* 6*d.*

Friends till Death. With Frontispiece. Fourteenth Thousand. Royal 16mo. Limp cloth, price 6*d.*

Two Christmas Stories. With Frontispiece. Eleventh Thousand. Royal 16mo. Limp cloth, price 6*d.*

Michel Lorio's Cross, and Left Alone. With Frontispiece. Seventh Thousand. Royal 16mo. Limp cloth, price 6*d.*

Old Transome. With Frontispiece. Ninth Thousand. Royal 16mo. Limp cloth, price 6*d.*

The Worth of a Baby, and how Apple-Tree Court was won. With Frontispiece. Ninth Thousand. Royal 16mo. Limp cloth, price 6*d.*

Hester Morley's Promise. 3 vols. Crown 8vo. Cloth.

The Doctor's Dilemma. 3 vols. Crown 8vo. Cloth.

STUMM (Lieut. Hugo), German Military Attaché to the Khivan Expedition.

Russia's advance Eastward. Based on the Official Reports of. Translated by Capt. C. E. H. VINCENT. With Map. Crown 8vo. Cloth, price 6*s.*

SULLY (James), M.A.

Sensation and Intuition. Demy 8vo. Cloth, price 10*s.* 6*d.*

Sunnyland Stories. By the Author of "Aunt Mary's Bran Pie." Illustrated. Small 8vo. Cloth, price 3*s.* 6*d.*

Tales of the Zenana. By the Author of "Pandurang Hari." 2 vols. Crown 8vo. Cloth, price 21*s.*

TAYLOR (Rev. J. W. A.), M.A.

Poems. Fcap. 8vo. Cloth, price 5*s.*

TAYLOR (Sir H.).

Edwin the Fair and Isaac Comnenus. A New Edition. Fcap. 8vo. Cloth, price 3*s.* 6*d.*

A Sicilian Summer and other Poems. A New Edition. Fcap. 8vo. Cloth, price 3*s.* 6*d.*

Philip Van Artevelde. A Dramatic Poem. A New Edition. Fcap. 8vo. Cloth, price 5*s.*

TAYLOR (Col. Meadows), C.S.I., M.R.I.A.

The Confessions of a Thug. Crown 8vo. Cloth, price 6*s.*

Tara: a Mahratta Tale. Crown 8vo. Cloth, price 6*s.*

TELFER (J. Buchan), F.R.G.S., Commander R.N.

The Crimea and Trans-Caucasia. With numerous Illustrations and Maps. 2 vols. Medium 8vo. Cloth, price 36*s.*

TENNYSON (Alfred).

Queen Mary. A Drama. New Edition. Crown 8vo. Cloth, price 6*s.*

TENNYSON (Alfred).

Cabinet Edition. Ten Volumes. Each with Frontispiece. Fcap. 8vo. Cloth. price 2*s.* 6*d.* each.

CABINET EDITION. 10 vols. Complete in handsome Ornamental Case. Price 28*s.*

TENNYSON (Alfred).

Author's Edition. Complete in Five Volumes. Post 8vo. Cloth gilt; or half-morocco, Roxburgh style.

VOL. I. **Early Poems, and English Idylls.** Price 6s.; Roxburgh, 7s. 6d.

VOL. II. **Locksley Hall, Lucretius, and other Poems.** Price 6s.; Roxburgh, 7s. 6d.

VOL. III. **The Idylls of the King** (*Complete*). Price 7s. 6d.; Roxburgh, 9s.

VOL. IV. **The Princess, and Maud.** Price 6s.; Roxburgh, 7s. 6d.

VOL. V. **Enoch Arden, and In Memoriam.** Price 6s.; Roxburgh, 7s. 6d.

TENNYSON (Alfred).

Original Editions.

Poems. Small 8vo. Cloth, price 6s.

Maud, and other Poems. Small 8vo. Cloth, price 3s. 6d.

The Princess. Small 8vo. Cloth, price 3s. 6d.

Idylls of the King. Small 8vo. Cloth, price 5s.

Idylls of the King. Complete. Small 8vo. Cloth, price 6s.

The Holy Grail, and other Poems. Small 8vo. Cloth, price 4s. 6d.

Gareth and Lynette. Small 8vo. Cloth, price 3s.

Enoch Arden, &c. Small 8vo. Cloth, price 3s. 6d.

Selections from the above Works. Super royal 16mo. Cloth, price 3s. 6d. Cloth gilt extra, price 4s.

Songs from the above Works. Super royal 16mo. Cloth extra, price 3s. 6d.

In Memoriam. Small 8vo. Cloth, price 4s.

TENNYSON (Alfred).

The Illustrated Edition. 1 vol. Large 8vo. Gilt extra, price 25s.

Library Edition. In 6 vols. Demy 8vo. Cloth, price 10s. 6d. each.

Pocket Volume Edition. 11 vols. In neat case, price 31s. 6d. Ditto, ditto. Extra cloth gilt, in case, price 35s.

Tennyson's Idylls of the King, and other Poems. Illustrated by Julia Margaret Cameron. 2 vols. Folio. Half-bound morocco, cloth sides, price £6 6s. each.

THOMAS (Moy).

A Fight for Life. With Frontispiece. Crown 8vo. Cloth, price 3s. 6d.

Thomasina.

A Novel. 2 vols. Crown 8vo. Cloth.

THOMPSON (Alice C.).

Preludes. A Volume of Poems. Illustrated by Elizabeth Thompson (Painter of "The Roll Call"). 8vo. Cloth, price 7s. 6d.

THOMPSON (Rev. A. S.).

Home Words for Wanderers. A Volume of Sermons. Crown 8vo. Cloth, price 6s.

Thoughts in Verse.

Small Crown 8vo. Cloth, price 1s. 6d.

THRING (Rev. Godfrey), B.A.

Hymns and Sacred Lyrics. Fcap. 8vo. Cloth, price 5s.

TODD (Herbert), M.A.

Arvan; or, The Story of the Sword. A Poem. Crown 8vo. Cloth, price 7s. 6d.

TRAHERNE (Mrs. A.).

The Romantic Annals of a Naval Family. A New and Cheaper Edition. Crown 8vo. Cloth, price 5s.

TRAVERS (Mar.).

The Spinsters of Blatchington. A Novel. 2 vols. Crown 8vo. Cloth.

TREMENHEERE (Lieut.-Gen. C. W.)

Missions in India: the System of Education in Government and Mission Schools contrasted. Demy 8vo. Sewed, price 2*s*.

TURNER (Rev. C. Tennyson).

Sonnets, Lyrics, and Translations. Crown 8vo. Cloth, price 4*s*. 6*d*.

TYNDALL (John), LL.D., F.R.S.

The Forms of Water in Clouds and Rivers, Ice and and Glaciers. With Twenty-five Illustrations. Sixth Edition. Crown 8vo. Cloth, price 5*s*.
Volume I. of The International Scientific Series.

UMBRA OXONIENSIS.

Results of the expostulation of the Right Honourable W. E. Gladstone, in their Relation to the Unity of Roman Catholicism. Large fcap. 8vo. Cloth, price 5*s*.

UPTON (Richard D.), Capt.

Newmarket and Arabia. An Examination of the Descent of Racers and Coursers. With Pedigrees and Frontispiece. Post 8vo. Cloth, price 9*s*.

VAMBERY (Prof. A.).

Bokhara: Its History and Conquest. Second Edition. Demy 8vo. Cloth, price 18*s*.

VAN BENEDEN (Mons.).

Animal Parasites and Messmates. With 83 Illustrations. Second Edition. Cloth, price 5*s*.
Volume XIX. of The International Scientific Series.

VANESSA.

By the Author of "Thomasina," &c. A Novel. 2 vols. Second Edition. Crown 8vo. Cloth.

VAUGHAN (Rev. C. J.), D.D.

Words of Hope from the Pulpit of the Temple Church. Third Edition. Crown 8vo. Cloth, price 5*s*.

The Solidity of true Religion, and other Sermons. Preached in London during the Election and Mission Week, February, 1874. Crown 8vo. Cloth, price 3*s*. 6*d*.

Forget Thine own People. An Appeal for Missions. Crown 8vo. Cloth, price 3*s*. 6*d*.

The Young Life equipping Itself For God's Service. Being Four Sermons Preached before the University of Cambridge, in November, 1872. Fourth Edition. Crown 8vo. Cloth, price 3*s*. 6*d*.

VINCENT (Capt. C. E. H.).

Elementary Military Geography, Reconnoitring, and Sketching. Compiled for Non-Commissioned Officers and Soldiers of all Arms. Square crown 8vo. Cloth, price 2*s*. 6*d*.

Vizcaya; or, Life in the Land of the Carlists at the Outbreak of the Insurrection, with some Account of the Iron Mines and other Characteristics of the Country. With a Map and Eight Illustrations. Crown 8vo. Cloth, price 9*s*.

VOGEL (Dr. Hermann).

The Chemical effects of Light and Photography, in their application to Art, Science, and Industry. The translation thoroughly revised. With 100 Illustrations, including some beautiful specimens of Photography. Third Edition. Crown 8vo. Cloth, price 5*s*.
Volume XV. of The International Scientific Series.

CURRENT COIN. By the Rev. H. R. HAWEIS, M.A., Author of "Speech in Season," "Thoughts for the Times," &c. Crown 8vo. Cloth.

Materialism—The Devil—Crime—Drunkenness—Pauperism—Emotion—Recreation—The Sabbath.

NOTES ON GENESIS. By the late Rev. F. W. ROBERTSON, M.A., Incumbent of Trinity Church, Brighton. Crown 8vo. Cloth.

SERMONS. Third Series. By the Rev. STOPFORD A. BROOKE, M.A., Chaplain in Ordinary to Her Majesty the Queen; and Minister at Bedford Chapel, Bloomsbury. Crown 8vo. Cloth.

RAYS FROM THE SOUTHERN CROSS—POEMS. By I. D. A. With Sixteen Full-page Illustrations by the Rev. P. WALSH. Crown 8vo. Cloth.

ANNUS AMORIS—SONNETS. By J. W. INCHBOLD. With a Specially Engraved Frontispiece. Fcap. 8vo. Cloth.

LAURELLA AND OTHER POEMS. By Dr. J. TODHUNTER.

NEW READINGS AND RENDERINGS OF SHAKESPEARE'S TRAGEDIES. By H. H. VAUGHAN. Demy 8vo. Cloth.

DAVID LLOYD'S LAST WILL. By HESBA STRETTON, Author of "Jessica's First Prayer," &c. Illustrated. Fcap. 8vo. Cloth.

SIR SPANGLE AND THE DINGY HEN. By LETITIA McCLINTOCK. Illustrated. Imperial 16mo. Cloth.

A STUDY FROM LIFE. By MISS M. DRUMMOND. Small crown 8vo. Cloth.

ELZEVIR PRESS:—PRINTED BY JOHN C. WILKINS,
9, CASTLE STREET, CHANCERY LANE.

Milton Keynes UK
Ingram Content Group UK Ltd.
UKHW021947021023
429820UK00005B/145